I0061048

# Ordinary Differential Equations and Applications II: With Maple Illustrations

Authored by

## Benjamin Oyediran Oyelami

*Department of Mathematics,*
*Plateau State University,*
*Bokkos, Nigeria*

*National Mathematical Centre,*
*Abuja, Nigeria*

*Baze University,*
*Abuja, Nigeria*

&

*University of Abuja,*
*Abuja, Nigeria*

**Ordinary Differential Equations and Applications II: with Maple Illustrations**

Author: Benjamin Oyediran Oyelami

ISBN (Online): 978-981-5313-86-4

ISBN (Print): 978-981-5313-87-1

ISBN (Paperback): 978-981-5313-88-8

© 2024, Bentham Books imprint.

Published by Bentham Science Publishers Pte. Ltd. Singapore. All Rights Reserved.

First published in 2024.

# BENTHAM SCIENCE PUBLISHERS LTD.
## End User License Agreement (for non-institutional, personal use)

This is an agreement between you and Bentham Science Publishers Ltd. Please read this License Agreement carefully before using the ebook/echapter/ejournal (**"Work"**). Your use of the Work constitutes your agreement to the terms and conditions set forth in this License Agreement. If you do not agree to these terms and conditions then you should not use the Work.

Bentham Science Publishers agrees to grant you a non-exclusive, non-transferable limited license to use the Work subject to and in accordance with the following terms and conditions. This License Agreement is for non-library, personal use only. For a library / institutional / multi user license in respect of the Work, please contact: permission@benthamscience.net.

## Usage Rules:

1. All rights reserved: The Work is the subject of copyright and Bentham Science Publishers either owns the Work (and the copyright in it) or is licensed to distribute the Work. You shall not copy, reproduce, modify, remove, delete, augment, add to, publish, transmit, sell, resell, create derivative works from, or in any way exploit the Work or make the Work available for others to do any of the same, in any form or by any means, in whole or in part, in each case without the prior written permission of Bentham Science Publishers, unless stated otherwise in this License Agreement.
2. You may download a copy of the Work on one occasion to one personal computer (including tablet, laptop, desktop, or other such devices). You may make one back-up copy of the Work to avoid losing it.
3. The unauthorised use or distribution of copyrighted or other proprietary content is illegal and could subject you to liability for substantial money damages. You will be liable for any damage resulting from your misuse of the Work or any violation of this License Agreement, including any infringement by you of copyrights or proprietary rights.

## Disclaimer:

Bentham Science Publishers does not guarantee that the information in the Work is error-free, or warrant that it will meet your requirements or that access to the Work will be uninterrupted or error-free. The Work is provided "as is" without warranty of any kind, either express or implied or statutory, including, without limitation, implied warranties of merchantability and fitness for a particular purpose. The entire risk as to the results and performance of the Work is assumed by you. No responsibility is assumed by Bentham Science Publishers, its staff, editors and/or authors for any injury and/or damage to persons or property as a matter of products liability, negligence or otherwise, or from any use or operation of any methods, products instruction, advertisements or ideas contained in the Work.

## Limitation of Liability:

In no event will Bentham Science Publishers, its staff, editors and/or authors, be liable for any damages, including, without limitation, special, incidental and/or consequential damages and/or damages for lost data and/or profits arising out of (whether directly or indirectly) the use or inability to use the Work. The entire liability of Bentham Science Publishers shall be limited to the amount actually paid by you for the Work.

## General:

1. Any dispute or claim arising out of or in connection with this License Agreement or the Work (including non-contractual disputes or claims) will be governed by and construed in accordance with the laws of Singapore. Each party agrees that the courts of the state of Singapore shall have exclusive jurisdiction to settle any dispute or claim arising out of or in connection with this License Agreement or the Work (including non-contractual disputes or claims).
2. Your rights under this License Agreement will automatically terminate without notice and without the

need for a court order if at any point you breach any terms of this License Agreement. In no event will any delay or failure by Bentham Science Publishers in enforcing your compliance with this License Agreement constitute a waiver of any of its rights.

3. You acknowledge that you have read this License Agreement, and agree to be bound by its terms and conditions. To the extent that any other terms and conditions presented on any website of Bentham Science Publishers conflict with, or are inconsistent with, the terms and conditions set out in this License Agreement, you acknowledge that the terms and conditions set out in this License Agreement shall prevail.

**Bentham Science Publishers Pte. Ltd.**
80 Robinson Road #02-00
Singapore 068898
Singapore
Email: subscriptions@benthamscience.net

**BENTHAM SCIENCE**

# CONTENTS

# FOREWORD

I strongly endorse this exceptional book on the subject of differential equations. It covers all aspects of the field. It has a solid theoretical foundation and an applied focus, with many practical examples. It demonstrates how to program them using Maple, which is a leading mathematical software; and finally, it demonstrates how to generate graphics that clearly represent the nature of solutions and provide deep insights into them. All of these aspects are essential in the use of differential equations in modern mathematics, science, and technology. Thus, the book is equally useful for mathematicians, scientists, and engineers. As engineers should have some understanding of the theory of differential equations, also mathematicians should be able to program and generate graphical results.

This volume is especially valuable because it presents all of these aspects in an integrated fashion. It is written by a true expert in the field, an experienced teacher who has also carried out significant applied research. As a master teacher, Dr. Oyelami presents the material in a simple, straightforward, easy-to-follow manner. As an expert researcher, he knows first-hand the power of differential equations as a modeling tool, and his love for the field is clearly visible. The volume is also comprehensive in its coverage, especially in the areas of differential equations of the greatest practical interest. The students who study this material will be thoroughly prepared for employment in technical fields that use differential equations for modeling purposes. Such a student will also find the book to be a valuable continuing reference, both for its clear theoretical presentations and its useful and generalizable computer codes.

**Christopher Thron**
Associate Professor of Mathematics,
Texas A&M University, Central Texas, USA

# ENDORSEMENT

I have thoroughly gone through this book, which can be considered to be a compendium of knowledge on Differential Equations at the Undergraduate levels in all ramifications. The book presents poignantly insightful views on quantitative and qualitative modeling, as it unleashes the tremendous power of differential equations techniques, with applications on current multifarious trends, including population dynamics, spread of viruses and diseases and neural networks.

This book places the generally neglected implementation aspects of mathematical results on the front burner, with special implementations on the platform of Maple. In the above regard, the contributions of this book are exceptional and unprecedented. In terms of scope and diversity, the reader will be surprised by the unfathomable depth of knowledge and broad horizon of the author on the mathematical modelling of continuous processes by the deft deployment of differential equations.

The book must be highly acclaimed for its balanced coverage of the theory, applications, and computational issues of differential equations and their solutions. It gives an effective exposition of differential equations and concepts with functional analytic support, as needed, with meticulously chosen examples, exercises and extensive use of Maple, currently regarded as the best mathematical software. This is the main thrust of the book, as it encompasses and emphasizes current trends of modern computational tools in enhancing the effectiveness of differential equations as an indispensable and core tool for modelling of processes that exist in the continuum.

On the other hand, the book reinforces the reader's understanding of ordinary differential equations, which, on the other hand, simulates and enhances the readers' interest and curiosity about the immense modelling possibilities on ordinary differential equations platforms.

This book vividly brings to the fore, the inconvertible fact that, for the most part, ordinary differential equations cannot be precluded in the modelling of real-life phenomena. This being an exceptionally well-crafted book with an abundance of realistic, well-researched examples, illustrations and exercises, will enliven discussions of ordinary differential equations, techniques and key modelling objectives that the reader will likely encounter in undergraduate courses and much more.

In view of the aforementioned attribute coupled with its lucid presentation, novelty of the abstract of each chapter and emphasis on digital implementations, this book deserves the highest recommendation. The book is a 'must-read 'and 'unputdownable'.

**Professor Ukwu  Chukwunenye**
Functional Differential Equations,
Control theory & Industrial Engineering Specialist,
Department of Mathematics,
University of Jos, Jos, Nigeria

# COMMENTS FROM RENOWNED SCIENTISTS

''I have gone through the whole book. It is simple, clear and easy to read and understand .I have no doubt in my mind that the book is a must for all students of Mathematics in Tertiary Institutions''

**Professor M.O. Ibrahim**
University of Ilorin
Ilorin and former President of Mathematics Association of Nigeria (MAN)

''The book will be a very good choice for both professionals across all fields of endeavours. The fact that the book does not assume familiarity with some basic mathematical concepts is an incentive in its appeal to those who have been out of school for some time. These qualities will increase its sellable quality in the market place as well as a recommendation to students on mathematical courses''.

**Professor Emeritus A.A. Asere**
Department of Mechanical Engineering,
Obafemi Awolowo University, Osun,
Nigeria

# PREFACE

Ordinary differential equations are powerful tools for modeling and analyzing complex phenomena in various fields. Understanding ODEs is essential for making accurate predictions, optimizing systems, and solving real-world problems. ODEs are vital tools in study involving climate change, population dynamics, economic growth, chemical reaction, resource management, epidemiological growth of diseases and pandemic and drug administration.

This textbook is an encyclopedia of techniques for finding solutions to ordinary differential equations. It was developed when lecturing students and researching at the Abubakar Tafawa Balewa University, Bauchi; Kaduna State University, Kaduna, Nigerian; Nile University, Abuja ; Plateau State University Bokkos, University of Abuja and Baze University Abuja all in Nigeria.

This book comprises nine chapters and it is on 'Vector valued ordinary differential equations and applications'. The Chapters are written bearing in mind beginners in the field of study who have little or no background on the course. This requirement is met by deployment of lucid and self-instructional language and utilization of scintillating examples throughout the book as well as illustration using Maple modeling and simulation software.

The first chapter contains preliminaries like set theory, topological concepts and the formulation of vector differential equations. The second chapter and the third chapter are on Linear differential equations in the linear space , basic concepts related to topological structures are discussed such structures are   Normed and Banach spaces as applicable to solutions of ordinary differential equations. The proof of existence and uniqueness of solution for initial value problems (IVP), the 'power house' of course finds its shape from fixed points.  Peano's existence theorem and Picard Lindelof theorem are exploited in no small measure. The fourth chapter is on solutions to matrix initial value problems.

The fifth chapter is about canonical transformation, a kind of transformation from scalar equations to vector equations. This chapter ends with the treatment of exponential matrices and estimation theory. The sixth is on Stability theory, Stability is a kind of graduation from continuous dependence on initial data localized to some finite interval of $E = (-\infty, +\infty)$ to more global generalized concepts. The seventh chapters examine the linear periodic systems with the

Floquent rule extensively utilized. Also treated in this chapter are stability of linear perturbed systems and applications to neural firing models, avian influenza, population models. The ninth chapter is on numerical solutions to ODEs and applications to some models respectively

Every part of the chapters in this textbook contains preambles without assuming students' familiarity with some basic mathematical concepts. Hence it will prove to be a valuable and supplementary textbook for other courses in Mathematics and Engineering.

**Benjamin Oyediran Oyelami**
Department of Mathematics, Plateau State University,
Bokkos, Nigeria, National Mathematical Centre, Abuja, Baze University,
Abuja, Nigeria

# ACKNOWLEDGEMENTS

My profound gratitude goes to Professor P. Smoczynski of the Department of Mathematics and Statistics, Simon Fraser University, Canada who first introduced me to Differential Equations and sustained my interest in the field. I am indebted to Professor Styr University of Botswana and late Professor Olaofe, University of Ibadan both of them taught me Numerical Analysis at undergraduate and postgraduate levels respectively. In ship of thanks are Prof. Ukwu Chukwunenye my Lecturer in the University of Jos Nigeria; Mr. Salam Mukaila a friend, late Professor M. Ibiejugba Koji State University Ayingba, Professor G. Abimbola and Professor M. O Ibrahim University of Ilorin, Nigeria; Professor Christopher Thron and Gwenda Lynn Anders, Texas A&M University-Central Texas, USA; late Professor P.C. Ram, and Professor M S.Sesay , Abubakar Tafawa Balewa University Bauchi, Nigeria.

Furthermore, I am grateful to: Professor S O Ale, National Mathematical Centre, Abuja, Nigeria; Professor A.A. Asere, Obafemi Awolowo University Ile-Ife Nigeria; Late Professor D.D. Bainov, Medical University Sofia Bulgaria; Professor Olusola Akinyele, Bowie State University, Maryland USA for their exposure to Impulsive differential equations and my mentor, Professor Emeritus Trench William Trinity College USA , and Professor R.A.T. Solarin and Professor Stephen Onah, the former Directors and Chief Executive of National Mathematical Centre(NMC), Abuja, Nigeria respectively. Professor Promise Mebine, the Present Director and Chief Executive of NMC. I am grateful to Professor Femi Taiwo Obafemi Awolowo University Ile-Ife. We are Co-Trainer for Maple Software across some Nigerian Universities. I am greatly indebted to the former Vice Chancellor, Plateau State University Bokkos, Professor Danjuma Sheni who through TETfund Research grant for preparation of this book. I am also grateful to Colleagues at the National Mathematical Centre, University of Abuja, Abuja and the Baze University all in Abuja, Nigeria.

I am grateful to the above academicians for their technical advice, thought provoking suggestions and eagle-eye proofreading of the whole manuscript. I am grateful to Mr. Anthony Oluloye of Tangier Company, who provided me with Maple 17, 18, 2015,2018,2019,2020,2021,2022,2023, MapleSim 7 and MapleSim 2023 gratis.

Finally, I must not forget my family especially my wife Keith E. Oyelami and my children; Moses, Ruth, Victoria, Miracle, David and Hannah for their contribution in making the publication of this book successful. I am grateful, God bless you all.

# DEDICATION

Dedicated to God Almighty, most merciful, the author of life, the giver of knowledge, wisdom and understanding. The procreator, sustainer, and annihilator of all life processes. God is the greatest problem solver who can solve a problem in an infinitely many ways.

# CHAPTER 1

# Vector-Valued Differential Equations and Related Analysis Concepts

**Abstract:** This chapter starts with the revision of basic concepts in real, complex, and functional analyses. Vector-valued differential equations are formulated and conditions for generating solution bases for the differential equations are stated.

**Keywords:** Basic concepts functional analyses, Solution bases, Vector-valued differential equations.

## INTRODUCTION

### What are Ordinary Differential Equations (ODEs)?

The branch of mathematics that studies equations involving derivatives of unknown functions is called differential equations. There are two classes of such equations that are classified according to the number of unknown variables involved. A differential equation is a relationship between an independent variable, x and dependent variable y, and one or more derivatives of y with respect to x. Differential equations with a single unknown variable are called ordinary differential equations (ODEs). ODEs find applications in mathematical physics, electrical engineering, and mechanical engineering, for example in the vibration of strings [5-8]

Ordinary Differential Equations (ODEs) are mathematical equations that describe how a quantity changes over time or space. They involve an unknown function and its derivatives, and are used to model a wide range of phenomena in science, engineering, economics, and other fields [1, 8].

ODE describes change over time or space .Typically involves rates of change (*e.g.*, velocity, acceleration) and can be linear or nonlinear. ODEs can be solved using various methods, including: Analytical methods, numerical methods, approximation methods (perturbation theory)

**Benjamin Oyediran Oyelami**
**All rights reserved-© 2024 Bentham Science Publishers**

In this textbook, efforts will be devoted to vector valued differential equations [5-8]. The system will be formulated in matrix form and solutions obtained in matrix form.

In most studies on scalar differential equations, the fundamental assumption made is that the solutions of the scalar equations exist and are uniquely determined. This may not be true in the general setting, therefore the need to establish the framework for existence and uniqueness of solutions of ODEs.   Theorems that guarantee the existence and uniqueness of solutions of ordinary differential equations [5-8] will be given in chapter three. Here we are considering the building blocks of tools for theorems on existence and uniqueness of solutions of ODES and associated analysis.

## PRELIMINARIES

Let us briefly review some of the familiar notions in the set theory relevant to our discussion in subsequent chapters.

## Open Sets

A set X is open, if there exists a neighborhood or ball that lies entirely in the set. Geometrically, a set X is open if we consider a sphere (ball) centered at $x_0$ with arbitrary radius $r_0$ which lies entirely inside the set. In set notation, we write $S(x_0) \subset X$ (See [3,4]

## Example 1.1

An arbitrary sphere in n-tuples Euclidean space:

1.     $$S(x_0, r_0) = \left\{ \begin{array}{l} (x_1, x_2, ..., x_n) \in E^n : (x_1 - x_{01})^2 + (x_2 - x_{02})^2 + (x_3 - x_{03})^2 + ... + (x_n - x_{0n})^2 < r_0 \\ , (x_{01}, x_{02}, ..., x_{0n}) \in E^n \end{array} \right\}$$

2.     $$S(x_0, r_0) = \{(x, y) \in E^2 : |x| < a, |y| < a, a \in E^1, a > 0\}$$

## Closed Set

A set X is closed if every open set in X lies in it, its boundary is inclusive.

## Example 1.2

A square $N(x_0, r_0)$ such that $N(x_0, r_0) = \{(x_1, x_2) \in E^2 : |x_1| \le a, |x_2| \le a, a > 0\}$.

A set could be open and closed simultaneously, examples are:

1. $$X = \{x \in E^1 : 0 \le x < 1\}$$

2. $$Y = \{x \in E^2 : |x_1| \le a, |x_2| < b, x = (x_1, x_2)\}$$

3. $$X_3 = \{x \in E^2 : (x_1 - x_{01})^2 + (x_2 - x_{02})^2 < 1\} \bigcap \{x \in E^2 : |(x_1, x_2)| \le 1\}.$$

## Bounded Set

X is a bounded set if there exists a positive constant M such that $|x| \le M$ for every $x \in X$.

## Compact Set

X is compact if any closed open subset whose union contains X has a finite subclass whose union also contains X. Heine-Borel theorem asserts that for a finite-dimensional Euclidean space, compactness is equivalent to closedness and boundedness, for example:

$$R(x, y) = \{(x, y) \in E^2 : |x| \le a, |y| \le b\} \text{ is compact in } E^2.$$

Reformulation of compactness in terms of open cover is given by William [4]. The equivalent definition to compactness is the Weistrass-Bolzano theorem, which is stated as follows: Any infinite sequence $\{x_n\}$ of $X$ has a subsequence $\{x_{nk}\}$ which converges to $x \in A$. This is often called subsequent compactness from a topological point [2,3].

## Connected Set

Set $X$ is connected if there exist two sets or points in X joined by an arbitrary line segment that lies entirely within X.

## Convex Set

Set $X$ is convex for $x_1, x_2 \in X$, there exists a constant $\lambda, 0 \le \lambda < 1$ such that $\lambda x_1 + (1-\lambda)x_2 \in X$. Examples are a plane (hyperplane) of any dimension, a half-plane or half space, a line segment, sphere, or any region in a plane including angle less than 45 degrees.

Further examples are: a unit sphere in $E^3$, and the 3D Euclidean space.

$$S = \{x = (x_1, x_2, x_3) \in E^3 : |x| \le 1\},$$

Let $x, y \in S$ then:

$$|\lambda x + (1-\lambda)y|$$
$$\le \lambda|x| + (1-\lambda)|y|$$
$$\lambda + 1 - \lambda = 1$$

$$. \to \lambda x + (1-\lambda)y \in S.$$

Another interesting feature of S is that it is closed and bounded in $E^2$, hence compact, by the Heine-Borel theorem. Any non-coplanar set is non-convex.

## FORMULATION OF VECTOR DIFFERENTIAL EQUATIONS

Let $t \in E^1$ and $\Omega$ be an open set in $E^{n+1}$ with element written as $(t,x)$: Let $f : \Omega \to E^n$ be a continous function written as $f \in C^0(\Omega, E^n)$ and $\bar{x} = \dfrac{dx}{dt}$.

A vector differential equation is a relation of the form:

$$\dot{x}(t) = f(t, x(t)) \tag{1.1}$$

where $x(t)$ is a solution to equation (1.1) on the interval $I \subset E, x(t)$ that is of class one defined on $I, (t,x) \in \Omega, t \in I$. We refer to $f$ as a vector field over $\Omega$. Now we are in a position to formulate a theorem that guarantees the existence and unique solution to ordinary differential equations. The Theorem shows that if W is a vector space, we claim that it is $n$ -dimensional, then we find a basis for W.

Let $\{x_\alpha\}_{\alpha=1,2,...,n}$ be a family of linearly independent solutions of the equation (1.1). We will show that it is the required basis of W. Suppose on the contrary $\{x_\alpha\}_{\alpha=1,2,...,n}$ is linearly dependent. Then there exist constants $C_\alpha$, not all the constants are zero

such that $\displaystyle\sum_{\alpha=1}^{n} C_\alpha x_\alpha = 0$

if the solutions of equation (1.1) satisfy the initial condition $x_\alpha(t_0) = e_j, j = 1,2,...,n$ , where $\{e_j\}$ is the $n$ -Euclidean basis of $E^n$ .

*i.e.*

$$e_j = \begin{pmatrix} 0 \\ 0 \\ . \\ . \\ 1 \\ 0 \end{pmatrix} \leftarrow \quad \text{j row where every element of } e_j, j = 1,2,...,n \text{ is zero except at j -th row,}$$

whose element is unity. Then:

$$\sum_{\alpha=1}^{n} C_\alpha x_\alpha(t_o) = \sum_{\alpha=1}^{n} C_\alpha e_\alpha = 0$$

$\to C_\alpha = 0, \forall \alpha \in \{1,2,...,n\}$, a contradiction of our assumption of linear dependence of $\{x_\alpha\}$ .

Finally, we will prove that $\{x_\alpha\}$ spans the solution space. This, in fact, is implied by the principle of superposition of solution such that $y = \displaystyle\sum_{\alpha=1}^{n} C_\alpha x_\alpha$ , the solution is unique. Therefore, $\{x_\alpha\}$ is the family of solutions and $\dim W = n$ . We claim that this solution is unique (Cf, Chapters 1 and 2).

## CONCLUSION

Basic topological structures such as open and closed sets, and other related concepts like compact sets, connected sets, and convex sets are presented in this chapter. It is worthy of note that topological structures are fundamental to understanding the theory of vector-valued differential equations. They are crucial to understanding how to generate a family of solutions to vector-valued differential equations. Most problems in real life from a differential equation perspective are multidimensional in nature, hence they are better treated using vector-valued differential equations. This chapter is a bridge between scalar differential equations and vector-valued differential equations associated with algebraic and topological structures.

## REFERENCES

[1]     C.T. Chen, *Introduction to Linear Systems Theory.*, 2nd ed Holt, Rinehart and Winston: New York, 1985.

[2]     Kreyszig. Erwin, *Advanced Engineering Mathematics.* John Wiley Publication: USA, 2000.

[3]     W. Rudin, *Real and Complex Analysis.* McGraw-Hill Book Company, 1986.

[4]     R. Wade William, *An Introduction to Analysis.*, 4th ed Pearson Prentice Hall Publication, 2010.

[5]     G. Deo Sadashiv, and V Ragavendra, *Ordinary Differential Equations..* Tata McGraw-Hill: India, 1980.

[6]     G. Deo Sadashiv, V Lakshimikantham, and V Ragavendra, *Textbook of Ordinary Differential Equations and Stability.* Tata McGraw-Hill: India, 1997.

[7]     F. Trench William, *Elementary Differential Equations with Boundary Value Problems.* Brooks/Cole-Thomson Learning: USA, 2013.

[8]     F. Trench William, *Elementary Differential Equations.* Brooks/Cole-Thomson Learning: USA, 2001.

# CHAPTER 2

# Differential Equations in the Linear Spaces

**Abstract:** Fundamental concepts in normed spaces are elucidated and linear systems are considered and applied to some problems.

**Keywords:** Gronwall's inequality, Linear systems, Norms, Normed spaces.

## INTRODUCTION

In this chapter, we introduce fundamental concepts useful for studying differential equations in linear space. The topological structure on spaces such as the Norm and Normed Space will be defined with some examples. We will also present Gronwall's inequality, the cornerstone of estimation theory in normed spaces [1, 6, 7, 8]. Many problems in differential equations are from Banach spaces, that is, complete normed spaces.

## PRELIMINARIES

A scalar function defined on the linear space $X(F)$ *i.e.,* $\|.\| : X \to E^+, E^+ = [0, +\infty)$ is called a norm on $X$ (see [1-8])

If the following conditions are satisfied:

1. $\quad\quad\quad\quad \|x\| \geq 0, \|x\| = 0$ if and only if $x = 0$, for $x \in V$

2. $\quad\quad\quad\quad\quad\quad \|x + y\| \leq \|x\| + \|y\|$

3. $\quad\quad\quad\quad\quad\quad \|\alpha x\| = |\alpha| \|x\|$

Property (ii) is called triangular inequality. This can be generalized to a finite number of arbitrary vectors in $V$ *i.e.,* $\|x_1 + x_2 + ... + x_n\| \leq \|x_1\| + \|x_2\| + ... + \|x_n\|$. At times, the triangular inequality is often referred to as Minskwoski 's inequality, if $\|.\|$ is defined on a metric space and satisfies all the properties of a norm, then the metric space is a normed space.

**Benjamin Oyediran Oyelami**
**All rights reserved-© 2024 Bentham Science Publishers**

The couple $[\|.\|, X]$ forms a normed space. A normed space with the additional structure of being linear (Chapters 5 and 6) is referred to as a normed linear space.

## Example 2.1

1. The space of continuous functions $C(E^n)$ defined in $E^n$ is equipped with a norm given by the pair $[C(E^n), \|.\|]$, forms a normed space [3, 4].
2. The vector space of bounded continuous functions from the interval $J$ to the set of positive real numbers. *i.e.*, with the norm, $\|f\| = \sup_{x \in E^n} |f| < \infty$ ,

$$X = \left\{ f \in C(J, E^+) : \|f\| = \sup_{t \in J} |f(t)|, J = [0, +\infty), \forall t \in J \right\}.$$

3. The Euclidean space $E^n$ endowed with any of the following norms forms a normed space:

$$\|x\| = \sum_{i=1}^{n} |x_i| \text{ for } x = (x_1, x_2, ..., x_n) \in E^n$$

$$\|x\| = \max_{1 \le i \le n} |x_i|, \|x\| = \left( \sum_{i=1}^{n} |x_i|^2 \right)^{\frac{1}{2}} \tag{2.1}$$

4. $L(J, Y)$ is the Banach space+ of all Lipchitzian functions in J strongly differentiable everywhere except for some finite number of points with range in the Banach space Y. $L(J, Y)$ is endowed with the sup norm $\|f\| = \sup_{f \in L(J,Y)} |f|$ .

+Readers familiar with metric spaces will quickly recognize that in the above equation, the three norms are equivalent in $E^n$ and as a matter of fact, this forms what is called topological isomorphism in the normed space. A linear space is a vector space, which, in addition, is linear (See William [4] and Chen [1]). Here it is not our interest to study topological structures in detail [2-4].

## Linear Systems

The general first-order n-dimensional linear system is a system of the form:

$$\dot{x}(t) = A(t)x(t) + f(t) \tag{2.2}$$

where $A(t)$ is the $n \times n$ matrix function of $t$ whose elements $[a_{ij}]_{i,j=1,\,2\,..\,n}$ are functions of $t \in E^1 = (-\infty, +\infty)$; $I$ is an open and connected in the sub-interval of $R$. If $f(t) = 0$, equation (2.2) is said to be a homogeneous linear system, otherwise, it is termed nonhomogeneous.

We remark that; if $(t_0, x_0) \in I \times \Omega$ *with* $|t_0| < \infty$. $\|x_0\| < \infty$ , $y(t_0) = x_0$ . Then there exists a unique solution $(t, x)$ passing through $(t_0, x(t_0))$ [3, 5]. This, would be justified in due course when the existence and uniqueness theorems are treated.

## Lemma 2.1 (Gronwall's inequality) (See [5-8])

Let $\alpha$ be a nonnegative real constant and let $\phi$ and $\beta$ be nonnegative and integrable on some interval $[a, b]$ such that:

$$\phi(t) \le \alpha + \int_{t_0}^{t} \beta(s)\phi(s)ds, a \le t \le b \tag{2.3}$$

For $a \le t_0 \le t \le b.$

Then:

$$\phi(t) \le \phi(a)\exp(\int_{t_0}^{t} \beta(s)ds)$$

$$\text{or } \phi(t) \le \alpha \exp(\int_{t_0}^{t} \beta(s)ds) \tag{2.4}$$

## Proof

$$\phi(t) \le \alpha + \int_{t_0}^{t} \phi(s)\beta(s)ds \tag{2.5}$$

Differentiating the equation (2.5) with respect to $t$, we get:

$$\dot{\phi}(t) \le \beta(t)\phi(t) \tag{2.6}$$

Separating the variables:

$$\frac{d\phi(t)}{\phi} \le \beta(t)$$

$$\int_{t_0}^{t} \frac{d\phi(s)}{\phi(s)} \le \int_{t_0}^{t} \beta(s)ds$$

$$\ln\left(\frac{\phi(t)}{\phi(a)}\right) \le \int_{a}^{t} \beta(s)\,ds$$

$$i.e.\, \phi(t) \le \phi(a)\exp\left(\int_{a}^{t} \beta(s)\,ds\right) \tag{2.7}$$

## Alternative Proof

Let $\psi(t) = \alpha + \int_{t_0}^{t} \beta(s)\psi(s)ds$ then, clearly $\phi(t) = \psi(t), \forall a \le t \le b$ and $\psi(t_0) = \alpha$.

Also, $\psi'(t) = \beta(t)\psi(t) \le \beta(t)\psi(t)$ a.e. on $[a,b]$ or $\psi'(t) = \beta(t)\psi(t) \le 0$ a.e. on $[a,b]$.

Clearly, the independent factor for the above differential inequality is $e^{-\int_{t_0}^{t}\beta(s)ds}$ thus

$\psi'(t)e^{-\int_{t_0}^{t}\beta(s)ds} - \beta(t)\psi(t)e^{-\int_{t_0}^{t}\beta(s)ds} \le 0$. The last inequality is exactly the same as

$$\frac{d}{dt}\left[\psi(t)e^{\int_{t_0}^{t}\beta(s)ds}\right]\leq 0 \text{ a.e. on}[a,b].$$ On integrating the inequality from $t_0$ to $t$, we get

$$\Rightarrow \psi(t)\leq \psi(t_0)e^{\int_{t_0}^{t}\beta(s)ds} \quad \text{or } \psi(t)\leq \alpha e^{\int_{t_0}^{t}\beta(s)ds} \quad \text{as desired.}$$

## Remark 2.1

i.  If $\phi$ and $\beta$ are nonnegative and continuous, then in particular, they are integrable and let the conditions in the hypothesis of Theorem hold. However, in the proof, the phrase "a.e. on $[a,b]$ " (a.e. is almost everywhere) may be omitted.

ii. $t_0$ may be replaced by $a$, with the condition preserved.

iii. The Proof strategy eliminates the 'division by zero' error.

## CONCLUSION

The topological structure the norm in a linear space form a normed linear Space. Complete normed linear spaces are Banach spaces; many problems in differential equations come from Banach spaces. We will also present Gronwall's inequality, the cornerstone of estimation theory in normed spaces.

## REFERENCES

[1]    C.T. Chen, *Introduction to Linear Systems Theory.*, 2nd ed Holt,Rinehart and Winston: New York, 1985.

[2]    Kreyszig Rudin, *Advanced Engineering.* John Wiley Publication: USA, 2000.

[3]    W. Erwin, *Real and Complex Analysis..* McGraw-Hill Book Company, 1986.

[4]    R. Wade William, *An introduction to Analysis.*, 4th ed Pearson Prentice Hall Publication, 2010.

[5]    G. Deo Sadashiv, and V Ragavendra, *Ordinary Differential Equations..* Tata McGraw-Hill: India, 1980.

[6]    G. Sadashiv, V Lakshmikantam, and V Ragavendra, *Textbook of Ordinary Differential Equations and Stability.* Tata McGraw-Hill: India, 1997.

[7]    F. Trench William, *Elementary Differential Equations with Boundary Value Problems.* Brooks/Cole-Thomson Learning: USA, 2000.

[8]    F. Trench William, *Elementary Differential Equations.* Brooks/Cole-Thomson Learning: USA, 2013.

# CHAPTER 3

# Fixed Point Theorems, Existence and Uniqueness of Solutions of Differential Equations

**Abstract:** In differential equations, one of the cornerstones of most theorems and their framework (hypotheses) is the existence and uniqueness theorem. We consider theorems that guarantee the existence and uniqueness of solutions of differential equations. We consider the Carathedory theorem, Peano existence theorem, Picard-Linderlof existence and uniqueness theorem, Brower and Schauder fixed point theorems. Picard successive approximation method is applied to establish the existence and uniqueness of solution to some initial value problems. Moreover, the conditions for the continuation of solutions from a given interval to an extended interval are also derived for ordinary differential equations.

**Keywords:** Existence, uniqueness, Differential equations, Carathedory theorem, Fixed point theorems, Peano existence theorem, Picard-Linderloft existence, uniqueness theorem, Picard successive approximation method.

## FIXED – POINT THEORY

Modern theorems on the existence and uniqueness of solutions to differential equations are from the so-called fixed theorems for which there are many versions [3,6-10]. Fixed point theory significantly utilizes some elements in functional analysis. This approach makes it a sophisticated and effective tool for solving differential equations.

Fixed point theory has a variety of applications widely used in both integral equations and operator theory [9-11]. Our goal or prime concern in this chapter is its application to initial value problems (IVP). Erwin [4], pp. 316 – 326) contains catalogues of applications of fixed-point theorems to integral equations and a systems of equations [14-15]. In particular, Garret and Gian [5] developed many iterative algorithms on fixed point theorems.

Furthermore, the following three types of fixed-point theorems are given: Schauder, Banach – Caccioppolis and Brower's fixed-point theorems. A concise statement of the theorems as was observed by Smart [14] is that every continuous mapping of a compact convex set to itself must have a fixed point. Besides, fixed point is a topological concept. It is fair for us to conclude from Smart's assertion that any closed interval $[a,b]$ in $E$ and a unit disc in $E^2$ must have a fixed-point property.

**Benjamin Oyediran Oyelami**
**All rights reserved-© 2024 Bentham Science Publishers**

Finally, the celebrated Picard-Linderlof existence and uniqueness theorem will be proved by a fixed-point theorem, to be specific, by Banach-Caccioppolis theorem. An alternative proof is found in Jack [7] where the proof was made by Schauder's fixed point theorem.

## Peano Existence Theorem [see [7], pp. 14 – 15]

If $f$ is continuous in a domain D, then for any initial data $(t_0, x_0)$, there exists at least one solution of the differential equation:

$$\dot{x}(t) = f(t, x(t)) \tag{3.1}$$

passing through $(t_0, x_0)$. Note if the assumption on $f$ in the piano existence theorem is satisfied, then the existence of infinitely many solutions to the equation (3.1) is guaranteed. We will see later on that for initial value problems, whenever their solutions exist, it is always uniquely determined if $f$ is bounded together with its first derivative or if it satisfies the Lipchitz condition.

## Method of Picard Successive Approximation

Solutions of differential equations can be approximated by sequences of points starting from the initial data. The successive approximated sequence of solutions $\{x_k\}$ converges to the actual solution of initial value problem (IVP) as $t \to \infty$, *i.e.* $x_k \to x$ as $t \to \infty$.

Picard presents an iteratively appealing method now termed Picard successive approximation method. This method overcomes the problem of obtaining solutions to differential equations via approximations. The method is due to J. Liouville and others in the early 1800s, but, often credited to E. Picard and hence it is called the Picard iterative method. E. Picard further developed the method in 1893 See [1] and [7].

The method is as follows:

Let $\dot{x}(t) = f(t, x(t)), x(t_0) = x_0$, where $t$ belongs to the half real line, $J = [0, +\infty)$ such that $f(t, x(t))$ is continuous and locally Lipchitzian in a rectangular domain

$R(\alpha,\beta) = \{(t,\alpha(t): |t-t_0| \le \alpha, |x(t)-x_0| \le \beta\}$ .Then that there exists a unique solution to equation (3.1) given by the successive approximation scheme:

$$x_k(t) = x_0 + \int_{t_0}^{t} f(s,x_{n-1}(s))ds, t \ge t_0 \text{ for } k \in \{0,1,2,...\}$$

The sequence $\{x_k\}$ converges uniformly to the solution $x(t)$ of equation (3.1) in the interval $J = [t_0,+\infty)$ .

## Lemma 3.1

Let $\qquad x_n(t) = x_0 + \int_{t_0}^{t} f(s,x_{n-1}(s))ds, t \ge t_0, m = \sup_{t \in [0,T]} |f(t,x(t))|$

Then:

$$\|x_1 - x_0\| \le m|t - t_0| \tag{3.2}$$

$$\| x_2 - x_1 \| \le \frac{L^2 m}{2!} |t - t_0| \tag{3.3}$$

L is Lipchitz constant. By induction on $k$ ,

$$\| x_{k+1} - x_k \| \le \frac{L^k m}{(k+1)!} |t - t_0|^{k+1} \tag{3.4}$$

Hence:

$$\sum_{0}^{\infty} \|x_{k+1} - x_k\| \le \frac{m}{L} \sum_{0}^{\infty} \frac{[L|t-t_0|]^{k+1}}{(k+1)!} = \frac{m}{L}[\exp L|t-t_0|-1]$$

$$\le \frac{M}{L}[\exp(\alpha L)-1] \tag{3.5}$$

$$\le \frac{M}{L}[\exp(L|t-t_0|)-1]$$

$\frac{M}{L}[\exp L|t-t_0|-1]$, in equation (3.5) is an estimation of upper error bound for $x$.

**Proof**

$$\|x_1-x_0\| = \sup\left|\int_{t_0}^{t}[f(s,x_1(s))-f(s,x_0(s))]ds\right|$$

$$\leq \int_{t_0}^{t}\|f(s,x_1(s))-f(s,x_0(s))\|ds$$

$$\leq \int_{t_0}^{t}L\|x_2-x_0\|ds$$

$$\leq \int_{t_0}^{t}Lm|s-t_0|ds \leq \frac{Lm|t-t_0|^2}{2!}$$

$$\|x_3-x_2\| \leq \frac{L^2m}{3!}|t-t_0|^3 \tag{3.6}$$

It can be established by induction on $k$ that:

$$\|x_{k+1}-x_k\| \leq \frac{L^k m}{(k+1)!}\sum_{0}^{\infty}\frac{[L|t-t_0|]^{k+1}}{(k+1)!}$$

$$\sum_{0}^{\infty}\|x_{k+1}-x_k\| \leq \frac{M}{L}\sum_{0}^{\infty}\frac{[L|t-t_0|]^{k+1}}{(k+1)!}$$

$$= \frac{M}{L}[e^{L(t-t_0)}-1] \leq \frac{M}{L}[e^{L\alpha}-1]$$

This estimation gives an upper bound for the error of approximation of the solution.

**Remark 3.1**

The series $\sum_{1}^{\infty}\|x_{k+1}-x_k\|$ is absolutely convergent, by comparison test, since $e^{\alpha t}$ is uniformly convergent.

**Theorem 3.1**

Let $f(t, x(t))$ satisfies the assumption imposed on it by equation (3.1). Then the sequence $\{x_k\}$ converges uniformly to the solution, $x(t)$ of the equation (3.1) in the interval $J = [0, +\infty)$.

**Proof**

(3.7)

$$x_m(t) = x_o + \sum_{k=0}^{m}\left(x_{k+1} - x_k\right)$$

(3.8)

$$x_e(t) = x_0 + \sum_{k=0}^{\ell}\left(x_{k+1} - x_k\right)$$

(3.9)

$$\left\|x_m - x_e\right\| \le \sum_{\ell=1}^{m}\left\|x_{k+1} - x_k\right\|$$

$$\le \frac{m}{L}\left[e^{\alpha L} - 1\right]$$

Let: $\lim_{m\to\infty} x_m = x(t)$                                                    (3.10)

Then:

$\left\|x - x_{(k)}\right\| \le \dfrac{m}{L}\left[e^{\alpha L} - 1\right]$ as $m \to \infty$ and by the continuity of norms, $x_k$ converges into

$x$ as $k \to \infty$ uniformly on $J = [0, +\infty)$ for $k \in \{1, 2, 3, ...\}$.

But:

$$\left\|x_k - x\right\| = \left\|\int_{to}^{t}\left[f(s, x_k(s,)) - f(s, x(s,))\right]ds\right\|$$

$$\le L\left\|x_k - x\right\| \to 0 \ as \ k \to \infty$$

Thus $\lim\limits_{k\to\infty} x_k(t) = x(t)$  *i.e.*    $x(t) = x_0 + \int_{t_0}^{t} f(s, x(s,))ds$  is the given solution to equation (3.1).

Finally, an alternative proof of the theorem can be made by the imposition of the stronger condition that: $f(t, x(t))$ is bounded by a non-increasing monotonic function in some interval of interest, provided uniqueness of the solution is guaranteed. The uniqueness theorem is our next theorem:

## Theorem 3.2 (Uniqueness Theorem)

Let:

$$\dot{x}(t) = f(t, x(t)), x(t_0) = x_0,  \tag{3.11}$$

where $f \in L_0(D)$. Then the space of the continuous function and locally Lipschizian function in a compact set D is elucidated. Let $U \subset D = J \times R, J = [0, +\infty), R \subset E^1$. The solution of the equation (3.11) passing through $[t_0, x_o)$ is unique.

## Proof

Let $x$ and $y$ be two solutions of equation (3.11) existing on interval $J = [o, \infty)$ passing through the initial point $(t_o, x_o)$ *i.e.* $y(t_o) = x_0$ and $x(t_o) = x_o$

Then:

$$x(t) = x_o + \int_{t_o}^{t} f(s, x(s))ds  \tag{3.12a}$$

$$y(t) = x_o + \int_{t_o}^{t} f(s, y(s))ds  \tag{3.12b}$$

Then:

$$\left\| x(t) - y(t) \right\| \leq \left\| \int_{t_0}^{t} [f(s, x(t)) - f(s, y(t))]ds \right\|$$

$$\leq L\int_{t_o}^{t}\| x(s)-y(s)\|\, ds \qquad (3.13)$$

Let $u(t) = x(t) - y(t)$

Then:

$$\| u(t)\| \leq 0 + L\int_{t_o}^{t}\| u(s)\|\, ds \qquad (3.14)$$

By Gronwall's Lemma:

$$\| u(t)\| \leq 0\exp(L\int_{t_o}^{t}\| u(s)\|\, ds) \text{ But } \|.\| \geq 0 \text{ by property of a norm.}$$

Hence:   $u(t) = x(t) - y(t) = 0$

This completes the proof.

## Orientation

We have discussed structures like linear spaces and norm spaces. Now, we will focus on another space which is a normed linear space with an additional topological property called completeness.

A normed linear space with an additional property of being complete is a Banach space. Banach space is named after a great Polish mathematician called S. Banach (1892 – 1945), who was the first to carry out a thorough investigation of the properties of these spaces.

A student of real analysis will recall that a space is said to be complete if every Cauchy (fundamental) sequence within the space is convergent in the space.

Let us refresh our memories with completeness concept:

## Definition 3.1

Let $X$ be a normed linear space. $X$ is said to be complete if given a Cauchy sequence $\{x_n\} \subset X$ such that $x_n \to x$ as $n \to \infty$, then $x_n \in X$.

We recall that, a sequence $\{x_n\}$ is Cauchy if given $\in > 0$, there exists $N(\in)$ (however large) such that:

$$\|x_m - x_n\| < \in \text{ for } m, n > N(\in) \tag{3.15}$$

A complete normed space is called a Banach space.

## Remark 3.2

1. Every finite dimensional subspace of $E^n$ is a Banach space.
2. Every sequence that is absolutely convergent in a Banach space, $X$ is also convergent.
3. The n-Euclidean space equipped with the Euclidean norm(See equation (2.1)) forms a Banach space.
4. The space of $L^p$, Lebesque measurable functions equipped with the Euclidean norm, $\|f\| = \left( \int_E |f|^p \right)^{\frac{1}{p}} < \infty$, where E is a measurable set, is a Banach.

5. The sequence space, $l^p$

$$l^p = \left\{ x = (x_1, x_2, ..., x_n) : \|x\| = \left( \sum_{i=1}^{n} |x_i|^p \right)^{1/p} \right\}$$

forms a Banach space.

6. The space of continuously differentiable functions $C^n(E)$, IS defined on measurable set E equipped the with the sup norm:

i.e., $\left\{ f \in C^n(E) : \|f\| = \sup_{x \in E} |f(x)| < \infty \right\}$ is another Banach space.

For further information on Banach spaces, refer to Boyce and Diprima [1], Huston and Pym [6], Yosida [15] and Kreyszig [6] for a deeper insight.

## Fixed Point Theorems

A vector $x$ in a normed linear space X is called a fixed point of $T$ , if it is a linear transformation (operator) such that $T : X \to X$ and $Tx = x$ for some $x \in X$ . That is, T sends a vector to itself.

## Contraction Mappings

A mapping $T : X \to X$ from a normed linear space $X$ onto itself is said to be a contraction map if there exists $0 \le \lambda \le 1$ such that:

$$\|Tx - Ty\| \le \lambda \|x - y\|, \tag{3.16}$$

for some $x, y \in X, \lambda$ is the contraction constant for $T$ .

## Geometric Interpretation

In $E^2$, the contraction constant can be interpreted as the slope of a straight line passing through the origin whose positive gradient is always less than 45° degrees or $\dfrac{\pi}{4}$ radian [14].

## Schauder Fixed Point Theorem

If X is a convex, a compact subset of a Banach space Y and $T : X \to X$ is continuous then, $t$ has a fixed point in $X$ . In other words, if T is completely continuous (sending a bounded set into a relative compact set) from $X$ onto itself, then there exists a fixed point for $T$ .

## Brower Fixed Point Theorem

Let $T : C^n \to C^n$ be a continuous mapping from closed unit ball in $E^n$ into itself. Then $T$ has a fixed point.

Note:

$$C^n = \left\{ x \in E^n : \|x\| \le 1 \right\} \tag{3.17}$$

Schauder fixed point theorem comes as a result of the generalization of Brower's theorem to the Banach space. Tychonof extended the concept to more general spaces [7]. Notice, as a matter of fact that $C^n$ defined in equation (3.17) is a convex and compact subset of $E^n$, which is a Banach space. The justification of generalization is authentic without a grit of salt.

## Theorem 3.3 (Banach – Cacciopolis Theorem)

Let F be a closed subset of a Banach space $X$ and T: $F \to F$ a contraction mapping. Then, there exists a unique fixed point $\bar{x} \varepsilon F$ of T. Suppose in addition $x_0$ is arbitrary in F, then:

$$\left\{ Tx_n : x_{n+1}, n \in \{0,1,2,...\} \right\} \to \bar{x} \text{ as } n \to \infty .$$

The estimation $\|x - x_n\| \le \dfrac{\lambda}{1-\lambda} \|x - x_0\|, n \in \{0,1,2,..\}$ is valid.

## Proof

Since $X$ is a Banach space, there exists $\{x_n\}, \{x_n\} \subset X$ which converges to $x \in X$.

*i.e.* given $\varepsilon > 0$, there exists a $N(\varepsilon)$ such that $\|x_n - x_m\| < \varepsilon$ implies $n > N(\varepsilon)$

$$\|x_m - x_n\| = \|Tx_{m-1} - Tx_{n-1}\|$$
$$\le \lambda \|x_{m-2} - x_{n-2}\|$$
$$\le \lambda^2 \|x_{m-3} - x_{n-3}\| \le ... \le \lambda^n \|x_{m-n} - x_0\|$$

$$\le \lambda^n \sup_{x_{m-n}, x_0 \in F} \|x_{m-n} - x_n\| \le \lambda^n diam(F) < \in$$

Where,

$$diamF \overset{def}{=} \sup_{x, y \in F} \|x - y\|$$

$$\lambda^n < (\varepsilon / diamF)$$

Then,
$$n > \frac{\ln(\varepsilon / diamF)}{\ln \lambda}, \text{for } 0 < \lambda < 1$$

Picking $N(\varepsilon) \geq \lim\limits_{\varepsilon \to 0}(\varepsilon / diamF), m, n \in N(\varepsilon)$ shows that $\{x_n\}$ is Cauchy sequence. But it is complete. Therefore, there is $\bar{x} \in B$, such that, $x_n \to x$. Since $F$ is a closed subset of Banach space, it follows that $\bar{x} \in F$.

Next, we must show that $\bar{x}$ is the fixed point of T, *i.e.* $T\bar{x} = \bar{x}$, then we will be done.

$$\|Tx_n - x_n\| = 0$$

$$\lim\limits_{n \to \infty} \| Tx_n - x_n \| = \| \lim\limits_{n \to \infty} Tx_n - \bar{x} \|$$

$\|T\bar{x} - \bar{x}\|$ by the continuity of T and $\|T\bar{x}\| = \bar{x}$.

## Uniqueness

Suppose $x, y \in F$, such that, $Tx = x, Ty = y$

$$\|x - y\| = \|Tx - Ty\| \leq \lambda \|x - y\|$$

Since $T$ is contraction by the dint of the theorem, $0 \leq \lambda < 1$,

$$(1 - \lambda)\|x - y\| \geq 0 \quad (\|.\| \geq 0)$$
$$\Rightarrow \|x - y\| = 0 \text{ i.e. } x = y$$

## Estimation

$$\|x_m - x_n\| = \|x_m - x_{m-1} + x_{m-1} + ... + x_{m-1} - x_n\|$$
$$\|x_m - x_{m-1}\| + \|x_{m-1} - x_{m-2}\| + ... + \|x_{m-1} - x_n\|$$
$$[\lambda^{m-1} + \lambda^{m-2} + .... + \lambda^n]\|x_1 - x_0\|$$
$$\leq \frac{\lambda^n}{1 - \lambda}\|x_1 - x_0\|$$

Let $\lim\limits_{m\to\infty} x_m = \bar{x}$ ,then $\left\| \bar{x} - x_n \right\| \le \dfrac{\lambda^n}{1-\lambda} \left\| x_1 - x_0 \right\|$

This ends the proof.

For applications of fixed-point theory to impulsive differential equations (See [9-13], fixed point theory is applied to a continuous function, $f(x)$, which is defined in a closed interval $[-1,+1]$.

A function $F(x) = x$ which is continuous on $[-1,+1]$ is well-defined. We notice the following: $F(+1) \le 0$ while $F(-1) \ge 0$ . It follows, therefore, by Intermediate value theorem that there exists $x_0 \in [-1,+1]$, such that $F(x_0) = 0$; $x_0$ is the fixed point of F(x) on $[-1,+1]$.

### Problem 3.1

1. Let $X$ be a normed space and $g(x), f(x)$ and $g(x) + f(x)$ be linear transformations on $X$ such that $x = g(x) + f(x)$ for some $x \in X$ . Show that $g(x)$ has a fixed point, $x_0 \in X$ . If $x_0$ is a zero of the function $f(x)$ .
2. Let $g(x)$ and $f(x)$ satisfy the assumptions on the problem above such that $x = 2g(x) - f(x)$ . Show that $g(x)$ cannot have a fixed point in $X$ for any non-zero vector $x \in X$ , if $x$ is a zero of $f(x)$ in $X$ .
3. Let $Tx = Ax$ be a linear transformation of Banach space $X = E^n$ , where $A$ is n-square matrix and $x$ is an n-column vector. Show that $T$ is a contraction map in $E^{\{n\}}$ if $\max \left| a_{ij} \right| \le K, 0 < K < 1$; furthermore, show that T has a fixed point if $A = I = $ the identity matrix of order n.

### CONTINUATION OF SOLUTION

The theory of existence and uniqueness guarantee the existence and uniqueness of solutions of initial value problems (IVP) in some interval of interest. Is it possible to have a solution of an IVP that may exist in an extended (larger) interval? This leads us to the concept of continuation of solutions.

## Definition 3.2

Consider the IVP $\dot{x} = f(t, x(t)), x(t_0) = x_0, t \in I = [0, +\infty)$ ; $\mathit{1}$ satisfying all assumptions for existence and uniqueness of solution of equation (3.1) in the given interval. A function $y(t)$ is a continuation of solution $x(t)$ of equation (3.1) to $I^*$, if $y(t)$ is defined in $I^*$, $x(t) = y(t)$ for $t \in I$ , $I \subset I^*$ and $y(t)$ satisfies the (IVP) on $I^*$.If the continuation just defined does not exist, then $I^*$ is the maximal interval of existence of solution $x(t)$. In other words, solution $x(t)$ cannot be extended beyond the natural boundary $\mathit{1}$ .

## Theorem 3.4

Let $f(t, x)$ be continuous and bounded for $(t, x) \in I \times \Omega$ and let the solution $x(t)$ exists on $I = [\alpha, \beta], \Omega \in E^n$ . Then the solution can be continued on I*,

$$\text{if } \lim_{t \to \alpha + o} x(t) \text{ and } \lim_{t \to \beta} x(t) \text{ exist.}$$

## Prove

that $x(t)$ is a solution of the equation, which implies that:

$$x(t) = x_0 + \int_{t_0}^{t} f(s, x(s)) ds, \alpha \le t_0 < t \le \beta .$$

$$\left\| x(t_2) - x(t_1) \right\| \le \int_{t_0}^{t} \left\| f(s, x(s)) \right\| ds \le m \left| t_2 - t_1 \right|$$

Since $f(t, x(t))$ is bounded, by our hypothesis, there exists a constant m such that:

$\left\| f(t, x) \right\| \le m, m > 0$. Thus $x(t_1) - x(t_2) \to 0$ as $t_1, t_2 \to \alpha + 0$ . This ends the proof. An appealing extension of the above Theorem is found in[3] and [7]; we refer the reader to this text for the proof of our next theorem:

## Theorem 3.5

Let $\varphi(t)$ and $\psi(x)$ be positive continuous functions on $\tau \leq t < \infty$ and $0 < x < \infty$, respectively, such that for any $A > 0$ $\lim\limits_{\beta \to \infty} \int\limits_A^B \dfrac{d\alpha}{\psi(x)} dx \leq +\infty$.

Then the solution of the IVP $\dfrac{dx}{dt} = f(t,x) = \phi(t)\psi(x)$, $n(\tau) = \varepsilon$, $\tau \geq t_o$ $\varepsilon > o$. can be continued (extended) to the right over the entire interval $\tau \leq t < \infty$ .

The solution of IVP:

$$\frac{dx}{dt} = t^2 e^{-x}, \; x(0) = 1.$$

## Picard –Linderloft Theorem

Consider initial value problem (IVP):

$$\dot{x}(t) = f(t, x(t)), x(0) = x_0$$

$f \in C^0(I \times R, E^n)$, where $R = \left\{ x \in E^n : \|x - x_0\| \leq a \right\}$, $I \subset E^1$. Assuming $f$ is locally Lipchitzian with respect to $x$ .

$$i.e. \|f(t,x) - f(t,y)\| \leq L\|x - y\|, L < \infty$$

$$\alpha = Min\left\{ \frac{k}{m}, \frac{1-\varepsilon}{L}, a \right\}, \; \varepsilon > 0$$

Then, the IVP has a unique solution passing through $(t_0, x_0)$ .

## Proof

Let $Tx = x_0 + \int\limits_{t_0}^t f(s, x(s))ds$, where $T : C(I, E^n) \to C(I, E^n), C(I, E^n)$, is a Banach space of continuous functions from $I$ into $E^n$ .

Define a subspace $A$ of $C(I, E^n)$ :

$$A = \{x \in C^n : \|Tx - x_0\| \le k, |t - t_0| < \infty, Tx(t_0) = x_0\} \not\subset C(I, E^n)$$

We claim that $T$ is contraction on $A$.

Proof of the Claim:

**Contraction**

Let $x, y \in A$ :   $\| Tx(s) - Ty(s) \| \le \int_{t_0}^{t} \| f(s, x(s)) - f(s, y(s)) \|$

$$\le L\alpha \| x - y \|$$
$$\le (1 - \varepsilon) \| x - y \|.$$

i.e   $\alpha \le \dfrac{(1 - \varepsilon)}{L}$   whenever $0 \le \varepsilon < 1$

Next, we show $T$ sends $C(I, E^n)$ to $C(I, E^n)$   i.e. $TC(I, E^n) \subset C(I, E^n)$.

$$\|Tx - x_0\| \le \left| \int_{t_0}^{t} f(s, x(s)) ds \right| \le \int_{t_0}^{t} \| f(s, x(s)) \| ds$$

$$\le M |t - t_0| \le \alpha M \le k$$

$$\Rightarrow \qquad \alpha \le \frac{M}{k}$$

$\alpha = Min\ [\dfrac{M}{k}, \dfrac{(1-\varepsilon)}{L}]$ which establishes that $TC(I, E^n) \subset C(I, E^n)$,

Since $T : C(I, E^n) \to C(I, E^n)$ is the contraction on the Banach space. $C(I, E^n)$ , Banach- Cacciopolis guarantees the existence of a unique fixed point of:

$$Tx_{n+1} = x_n + \int_{t_o}^{t} f(s, x_n(s)) ds,\ t \ge t_o$$

Let, $\lim_{n\to\infty} x_n = \alpha \; \lim_{n\to\infty} Tx_{n+1} = T\lim_{n\to\infty} x_{n+1} = Tx = x_o + \int_{t_o}^{t} f(s,x(s))ds = x(t)$.

This is the unique solution which coincides with the fixed points of T.

### Remark 3.3

This could as well be proved by Piano existence theorem and the uniqueness theorem.

### Example 3.1 (Picard's successive approximation)

Find the approximate solution to:

$$\begin{cases} \dot{x}(t) = tx(t) \\ \quad x(0) = 1 \end{cases}$$

using the successive approximation method.

### Solution

$f(t,x(t)) = tx(t), x(0) = 1$.

By Picard's successive approximation method,

$$x_{n+1} = x_n + \int_{t_0}^{t} f(s,x_n(s))ds$$

$$= x_0 + \int_{t_0}^{t} sx_n(s)ds$$

When,           $x_0 = 1, x_1 = x_0 + \int_{t_0}^{t} sds = 1 + \frac{t^2}{2}$.

$$x_2 = x_1 + \int_{t_0}^{t} x_1 ds = 1 + \frac{t^2}{2} + \int_{t_0}^{t} s(1 + \frac{s^2}{2})ds$$

$$= 1 + \frac{t^2}{2} + (\frac{s^2}{2} + \frac{s^4}{2.4})\Big|_0^t = 1 + \frac{t^2}{2} + \frac{t^2}{2} + \frac{t^4}{4.8}$$

$$\text{As } k \to \infty, x_k \to x_0 e^{\frac{t^2}{2}} = e^{\frac{t^2}{2}}$$

## Maple Example Fixed Point Iterations

> $$f := (x, y) \to 2 \cdot x + x^2 + y^2;$$

$$f := (x, y) \mapsto 2 \cdot x + x^2 + y^2$$

> $$a := 0;$$

$$a := 0$$

> $$\phi_0 := 0;$$

$$\phi_0 := 0$$

> $$N := 4;$$

$$N := 4$$

>

```
for k from 0 to N − 1 do
  φ_{k+1} := unapply( φ_0 + ∫_a^x f(t, φ_k(t)) dt, x );
  print(nprintf("Iterate number %d:", k + 1)) :
  print( φ_{k+1}(x) ); end do:
```

*Iterate number 1:*

$$\frac{1}{3} x^3 + x^2$$

*Iterate number 2:*

$$\frac{1}{63}x^7 + \frac{1}{9}x^6 + \frac{1}{5}x^5 + \frac{1}{3}x^3 + x^2$$

*Iterate number 3:*

$$\frac{1}{59535}x^{15} + \frac{1}{3969}x^{14} + \frac{53}{36855}x^{13} + \frac{1}{270}x^{12} + \frac{239}{51975}x^{11} + \frac{2}{189}x^{10} + \frac{16}{405}x^9$$

$$+ \frac{1}{20}x^8 + \frac{1}{63}x^7 + \frac{1}{9}x^6 + \frac{1}{5}x^5 + \frac{1}{3}x^3 + x^2$$

*Iterate number 4:*

$$\frac{1}{109876902975}x^{31} + \frac{13231}{10614240}x^{16} + \frac{96110851}{173675502000}x^{17} + \frac{7025719}{76621545000}x^{20}$$

$$+ \frac{4797722}{21837140325}x^{19} + \frac{7631}{19646550}x^{18} + \frac{7309987}{160905244500}x^{21} + \frac{2925773}{117997179300}x^{22}$$

$$+ \frac{1057035773}{101772567146250}x^{23} + \frac{141341}{39405366000}x^{24} + \frac{23}{758475900}x^{28}$$

$$+ \frac{2138222}{14119435204875}x^{27} + \frac{107203}{209176817850}x^{26} + \frac{16220423}{12067893337500}x^{25}$$

$$+ \frac{1}{3544416225}x^{30} + \frac{1717}{445414972275}x^{29} + \frac{16}{405}x^9 + \frac{1}{20}x^8 + \frac{827}{291060}x^{14}$$

$$+ \frac{563}{110565}x^{13} + \frac{127}{9720}x^{12} + \frac{1423}{103950}x^{11} + \frac{2}{189}x^{10} + \frac{82163}{32744250}x^{15} + \frac{1}{5}x^5$$

$$+ \frac{1}{63}x^7 + \frac{1}{9}x^6 + x^2 + \frac{1}{3}x^3$$

> $plot\left(\phi_{k+1}(x), x = 0..1\right);$

Fig. (**3.1**) is **a** graph for Fixed Point Iterations for function $f(x, y) = 2x + x^2 + y^2$

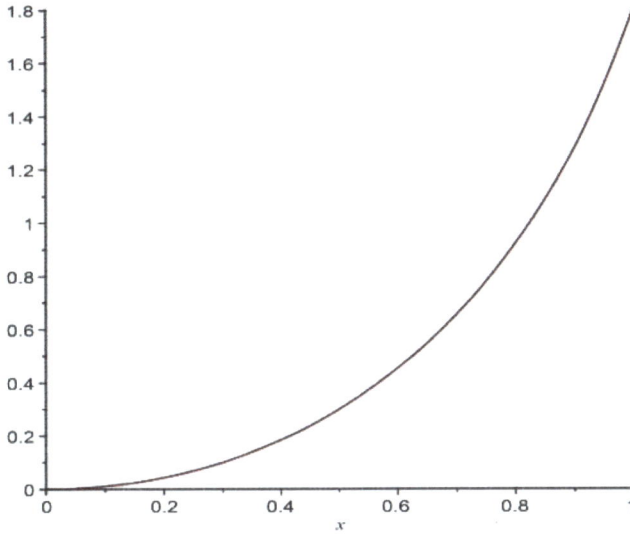

**Fig. (3.1).** Graph for Fixed Point Iterations for $f(x, y) = 2x + x^2 + y^2$.

**In the following Example, using maple, we will consider** Picard's successive approximation for the problem in Example 3.1. The graph for the tenth iteration is plotted in Fig. **(3.2)**.

> $\qquad\qquad\qquad\qquad f(t, x) := t \cdot x;$

$\qquad\qquad\qquad\qquad f := (t, x) \mapsto x \cdot t$

> $\qquad\qquad\qquad\qquad \text{phi}[0] := 1;$

$\qquad\qquad\qquad\qquad \phi_0 := 1$

> $\qquad\qquad\qquad\qquad N := 10;$

$\qquad\qquad\qquad\qquad N := 10$

> **for** $k$ **from** $0$ **to** $N - 1$ **do** $\text{phi}[k + 1] := unapply\left(\text{phi}[0] + \int_0^x f(t, \text{phi}[k](t))\, dt, x\right)$ **end do;**

$$\phi_1 := x \mapsto 1 + \frac{x^2}{2}$$

$$\phi_2 := x \mapsto 1 + \frac{1}{8} \cdot x^4 + \frac{1}{2} \cdot x^2$$

$$\phi_3 := x \mapsto 1 + \frac{1}{48} \cdot x^6 + \frac{1}{8} \cdot x^4 + \frac{1}{2} \cdot x^2$$

$$\phi_4 := x \mapsto 1 + \frac{1}{384} \cdot x^8 + \frac{1}{48} \cdot x^6 + \frac{1}{8} \cdot x^4 + \frac{1}{2} \cdot x^2$$

$$\phi_5 := x \mapsto 1 + \frac{1}{3840} \cdot x^{10} + \frac{1}{384} \cdot x^8 + \frac{1}{48} \cdot x^6 + \frac{1}{8} \cdot x^4 + \frac{1}{2} \cdot x^2$$

$$\phi_7 := x \mapsto 1 + \frac{1}{645120} \cdot x^{14} + \frac{1}{46080} \cdot x^{12} + \frac{1}{3840} \cdot x^{10} + \frac{1}{384} \cdot x^8 + \frac{1}{48} \cdot x^6 + \frac{1}{8} \cdot x^4 + \frac{1}{2} \cdot x^2$$

$$\phi_8 := x \mapsto 1 + \frac{1}{10321920} \cdot x^{16} + \frac{1}{645120} \cdot x^{14} + \frac{1}{46080} \cdot x^{12} + \frac{1}{3840} \cdot x^{10} + \frac{1}{384} \cdot x^8 + \frac{1}{48} \cdot x^6 + \frac{1}{8} \cdot x^4 + \frac{1}{2} \cdot x^2$$

$$\phi_9 := x \mapsto 1 + \frac{1}{185794560} \cdot x^{18} + \frac{1}{10321920} \cdot x^{16} + \frac{1}{645120} \cdot x^{14} + \frac{1}{46080} \cdot x^{12} + \frac{1}{3840} \cdot x^{10} + \frac{1}{384} \cdot x^8 + \frac{1}{48} \cdot x^6 + \frac{1}{8} \cdot x^4 + \frac{1}{2} \cdot x^2$$

$$\phi_{10} := x \mapsto 1 + \frac{1}{3715891200} \cdot x^{20} + \frac{1}{185794560} \cdot x^{18} + \frac{1}{10321920} \cdot x^{16} + \frac{1}{645120} \cdot x^{14} + \frac{1}{46080} \cdot x^{12} + \frac{1}{3840} \cdot x^{10} + \frac{1}{384} \cdot x^8 + \frac{1}{48} \cdot x^6 + \frac{1}{8} \cdot x^4 + \frac{1}{2} \cdot x^2$$

> *plot( (4) )*

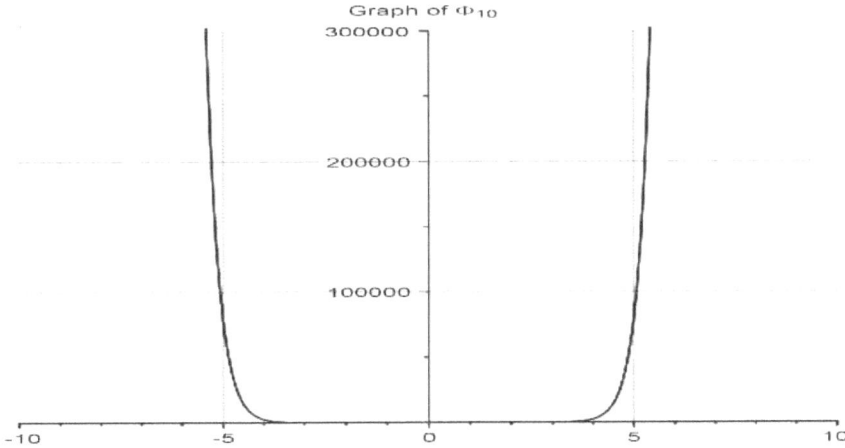

**Fig. (3.2).** The term for Picard's successive approximation for the problem in Example 3.1.

## Problems 3.2

1. Let   E   be   a   Banach   space   and   consider   the   ball:
$\sum' = \{x \in E, \|x\| < P, P > 0\}$   Assuming $f : \sum' \to \sum'$ is a contraction
mapping such that $\|f(\theta)\| \le p(1 - p)$ where $\theta$ is the zero element of $\sum'$ .
Show that $f(\sum') \subset \sum'$ . Hence, show that there exists a unique fixed point
$x^* \in \sum'$ .

2. Let $K$ be a closed convex subset of a Banach space $E$. Assume $f$   and $g$ are
   mappings from $K$ into $E$ such that the following conditions are satisfied:
   a.   $f(x) + g(x) \in E$ for $x, u \in K$
   b.   $f$ is contraction mapping
   c.   $g$ is continuous and carries any bounded set into a relatively compact set.

Prove that there exists at least one fixed point of $g(x) + f(x)$ in $K$ .

1. Let $E$ be a Banach space and consider a linear operator $A : E \to E$. Assume that
   $f : E \to E$ is a mapping, satisfying a Lipchitz condition. Discuss the existence
   of solutions of the equation.

$$\dot{x} = Ax + f(x)$$

If T is a contraction mapping of a Banach space V into itself, show that the equation $Tf - f = g$ has a unique solution $f$ for each $g$ in V. Also, show that T − I and $(T - I)^{-1}$ are uniformly continuous.

## CONCLUSION

Fixed point theorem is elucidated together with some varieties of it such as the piano existence theorem, and Picard successive approximation method. Other methods considered are contraction mapping, Schrader, Brower, Banach-Caccioppoli and Picard-Lindelof fixed theorems. The existence and uniqueness of solutions to initial value problems are established using some of these fixed point theorems mentioned and continuation of solutions.

## REFERENCES

[1]    W.E. Boyce, and R.C. Diprima, *Elementary Differential Equations and Boundary Value problems.,* 7th ed John Wiley and Son, 1977.

[2]    H. Chirgwin Brain, and Plumpton Charles, *Advanced Theoretical Mechanics: A Course of Mathematics..* Pergamon Press Ltd, 1961.

[3]    A. Coddington Earl, and Levinson Norman, *Theory of Ordinary Differential Equations..* McGraw Hill, 1955.

[4]    Kreyszig Erwin , *Advanced Engineering.* John Wiley Publication: USA, 2000.

[5]    Birkhoff Garrett , and Rota Gian-Carlo, *Gian-Carlo Rota, Ordinary Differential Equations.* Wiley, 1989.

[6]    V.C Huston, and JJ Pym , *Application of Functional Analysis and Operator Theory.* Academic Press London: New-York / Toronto, 1989.

[7]    K. Jack, *Hale "Ordinary Differential Equations.* Wiley-Interscience: New York, London, Sydney, Toronto, 1969.

[8]    E. Kreyszig, *Introduction to Functional Analysis with Application.* Wiley Indian Publication, 2007.

[9]    V. Lasksmikantham, D.D Bainov , and P.S. Simeonov , *Theory of Impulsive Differential Equations.* World Scientific Publications: Singapore, 1989.

[10]   B.O. Oyelami, S.O. Ale, and M.S. Sesay, "On existence", *J. Niger. Math. Soc.,* vol. 21, pp. 13-25, 2002.

[11]   B.O. Oyelami, S.O. Ale, and M.S. Sesay, "On existence", *Abuja Conference in Ordinary Differential Equations,* 2000, pp 101-117.

[12]   B.O. Oyelami, S.O. Ale, and M.S. Sesay, "Impulsive cone value integrodifferential and differential inequalities", *Electron. J. Differ. Equ.,* vol. 66, pp. 1-14, 2005.

[13]   B.O. Oyelami, "On existence of solution, oscillation and non-oscillation properties of delay equations containing maximum", *Acta Appl. Math. J.,* 2010, 109, 683-701.
       http://dx.doi.org/10.1007/s10440-008-9340-1

[14]    R.D. Smart, *Fixed point* . Cambridge University Press, 1974.

[15]    Kosasa. Yosida, *Functional Analysis,* 6th Springer-Verlag.: Berlin, 1980.

# CHAPTER 4

# Matrix Solution to Initial Value Problems

**Abstract:** In this chapter, we will consider methods for estimating the norm of a matrix and matrix exponents. The conditions for the existence and uniqueness of solutions are considered for ordinary differential equations using the Lipchitz conditions. Adjoint systems are revisited together with the application of the Carathedory theorem to some selected problems.

**Keywords:** Adjoint systems, Carathedory theorem, Lipchitz conditions, Matrix exponents, Ordinary differential equations, Uniqueness.

## INTRODUCTION

We extend the idea of the norm of a vector to a matrix. Matrix exponentials of a vector-valued differential equation (VDEs) are found in this chapter. Fundamental matrix solutions will be obtained for VDEs together with their corresponding adjoint systems using a vector version of variation of constant parameters.

Consider:

$$\dot{x}(t) = A(t)x(t) \tag{4.1}$$

On $J = [\alpha, \beta]$, the matrix $A(t)$ plays a central role. It will be useful to evaluate a norm for a matrix [1-5].

We define such a norm and illustrate its usefulness in estimation and continuous dependence of the solution of equation (4.1) on the initial data $x(t_0) = x_0$.

## NORMS FOR MATRICES

### Definition 4.1

The norm of matrix A with elements $[a_{ij}]$ will be defined as:

$$\|A\| = \max_{1 \le j \le n} \sum_{i}^{n} |a_{ij}| \tag{4.2}$$

**Benjamin Oyediran Oyelami**
**All rights reserved-© 2024 Bentham Science Publishers**

*i.e.* the maximum of the sum of the absolute value of elements in each column. Many other definitions are available, for example, an alternative definition analogous to the norm of an operator is given as:

$$\|A\| = \limsup\nolimits_{\|x\| \le 1} \|Ax\|$$

$$= \limsup\nolimits_{\|x\| \ne 0} \frac{\|Ax\|}{\|x\|} \tag{4.3}$$

$$= \inf \left\{ k : \|Ax\| \le \|x\| \right\}$$

*i.e.* the best upper bound for $Ax$.

At times, in some texts, one finds it being defined as:

$$\|A\| = \sum_{j=1}^{n} \sum_{i=1}^{n} |a_{ij}| \tag{4.4}$$

*i.e.* the sum of the absolute entries of matrix $[a_{ij}]$. It must be mentioned that each of these norms is equivalent to one another, the square of the norm $\|A\|$, *i.e.* $\|A\|^2$ in the equation. (4.3) is the maximum of each value of matrix A.

**Norm of a Vector**

The norm of a vector with components $e_1, e_2, \ldots, e_n$

is defined by
$$\|E\| = \sum_{i=1}^{n} |e_i|$$

Note,        $\min\limits_{i,j} \|a_{ij}\| \|x\| \le \|Ax\| \le \max\limits_{i,j} \|a_{ij}\| \|x\|$  (see [1,7,9-11])

Observe that $\|I\| = 1$.

**Properties of Matrix Norm**

Let A and B be square matrices and I is a column n-vector

Such that:

i.   $\|A\| \geq 0$, $\|A\| = 0 \Rightarrow A = [0]$, zero matrix

ii.   $\|A + B\| \leq \|A\| + \|B\|$ (Triangular inequality)

iii.   $\|\alpha A\| = |\alpha| \|A\|$ For every $\alpha$ complex number.

iv.   $\|A\varepsilon\| \leq \|A\| \|\varepsilon\|$

The reader will quickly notice properties; (i) through (iii) as those of a norm.

Properties together with the vector space of square matrices constitutes a normed space.

**Example 4.1**

Prove the above-stated properties of the norm of a matrix.

1.   $\|A\| = \max\limits_{j \in \{1,2,3,\dots,n\}} \sum\limits_{i=1}^{n} |a_{ij}|$. Since $|a_{ij}| \geq 0$, $|a_{ij}| = 0$ if and only if $a_{ij} = 0$. Therefore,

$$\|A\| \geq 0, \|A\| = 0 \text{ if and only if } A = [0]$$

$$\|A + B\| = \max\limits_{1 \leq j \leq n} \sum\limits_{i=1}^{n} |a_{ij} + b_{ij}|$$

$$\leq \max\limits_{i,j} \sum\limits_{i,j} \left( |a_{ij}| + |b_{ij}| \right)$$

$$= \|A\| + \|B\|$$

2.        $\|AE\| = \max\limits_{\substack{1 \leq j \leq n \\ 1 \leq i \leq n}} \sum |a_{ij}| \|E_j\| = \|A\| \|E\|$

In the same vein,

$$\|AB\| = \max \sum\limits_{j=1}^{n} \sum\limits_{i=1}^{n} |a_{ij} b_{ij}| \leq \max \sum\limits_{j=1}^{n} \sum\limits_{i=1}^{n} |a_{ij}| |b_{ij}|$$

$$\leq \|A\| \|B\|$$

## Lemma 4.1

If A and B are two square matrices, then:

$$\|A\| - \|B\| \leq \|A - B\| \qquad (4.5)$$

## Proof

$$\|A\| - \|B\| \leq \|A - B\| + \|B\| \qquad (4.6)$$

*i.e.*

$$\|A\| - \|B\| \leq \|A - B\|$$

Similarly:

$$\|B\| = \|B - A + A\| \leq \|B - A\| + \|A\|$$

$$i.e. \quad \|B\| - \|A\| \leq \|B - A\| \qquad (4.7)$$

(4.6)   implies (4.7):

$$\|A\| - \|B\| \leq \|A - B\|$$

## Examples 4.2

Consider the equations:

$$\dot{x}_1 = -\sin t x_1 + 4$$

$$\dot{x}_2 = -x_1 + 2t x_2 - x_3 + e^t$$

$$\dot{x}_3 = 3x_1 \cos t x_1 + x_2 + \frac{2}{t} x_3 - st$$

$$0 \leq t \leq 1$$

## Solution

$$\|A(t)\| = \max_{0 \le t \le 1} \left\{ \left|-\sin t\right|, \left|-1\right| + \left|2t\right| + \left|-1\right|, \left|3\cos t\right| + 1 + \left|\frac{2}{t}\right| \right\}$$

$$\le \max\{1, 1+2+1, 3+1+2\}$$

$$= \max\{1, 4, 6\} = 6$$

$$h(t) = \begin{pmatrix} 4 \\ e^t \\ -st \end{pmatrix}, \|h(t)\| = \max_{0 \le t, s \le 1} \left( \left|4\right| + \left|e^t\right| + \left|-st\right| \right) = 4 + e + 1 = 4 + e.$$

## Matrix Exponentials

Consider the equation:

$$\left. \begin{array}{l} \dot{x} = Ax + h(t) \\ x(t_0) = x_0 \end{array} \right\} \tag{4.8}$$

A study of the behavior of solutions of ordinary differential equations is often achieved by the investigation of properties of the exponential matrix, otherwise called the principal matrix.

Let us introduce the exponential matrix $e^A$ given by:

$$e^A = I + A + \frac{A^2}{2!} + \dots \tag{4.9}$$

If $X(t)$ is any fundamental matrix solution of the equation (4.1) thus, the general solution is of the form:

$$X(t) = \phi(t)C \tag{4.10}$$

C is an arbitrary constant vector.

**Proof**

Let:

$$x(t) = \phi(t)C$$

Then:

$$\dot{x}(t) = A(t)\phi(t)C$$
$$= A(t)x(t)$$

Since $\dot{x}(t) = A(t)x(t)$

$x(t)$ is a solution of (10.5) imposing the initial condition, we can determine C. Thus:

$$x(t_0) = \phi(t)C \Rightarrow \phi^{-1}(t_0)x_0 = C \tag{4.11}$$

$\phi(t)$ is the principal matrix solution such that:

$\phi(t_0)\phi^{-1}(t_0) = I$ =Identity matrix.

In the same vein, the solution of a nonhomogeneous system can be obtained.

$$\dot{x}(t) = A(t)x(t) + h(t) \tag{4.12}$$

Assuming $h(t)$ is integrable and can readily determine the variation in constant parameters, but this time only for vector differential equations.

$$x(t) = \phi(t)\phi^{-1}(t_0)x_0 + \phi(t)\int_{t_0}^{t}\phi^{-1}(s)h(s)ds, t_0 \geq 0$$

Define:

$$\phi(t,s) = \phi(t)\phi^{-1}(s) \text{ (See [7])}$$

Thus:

$$\phi(t,s) = \phi(t,\tau)\phi(\tau,s) \tag{4.13}$$

Then:

$$x(t) = \phi(t,t_0)x_0 + \int_{t_0}^{t} \phi(t,s)h(s)ds \qquad\qquad \textbf{(4.14)}$$

Method of variation of constant parameter, the solution of non-homogenous system is:

$$x(t) = \phi(t)C(t)$$

$C(t)$ is an undetermined vector.

Thus:

$$x = \phi(t)c + \phi(t)C(t)$$

But , $x = A(t)x(x) + h(t)$ hence:

$$\phi(t)C + \phi(t)c(t) = A(t)\phi c + h(t)$$
$$\Rightarrow$$
$$A\phi c + \phi c = A\phi c + h$$
$$\phi c = h$$
$$C^* = \phi^{-1}(t)h(t)$$
$$C(t) = \int \phi^{-1}(s)h(s)ds$$

Hence:

$$X_p = \phi(t) \int \phi^{-1}(s)h(s)ds = \int \phi(t,s)h(s)ds$$

$$X(t) = \phi(t)C + \int \phi(t,s)h(s)ds$$

$$= \phi(t)\phi^{-1}(to)x_o + \int \phi(t,s)h(s)ds \qquad\qquad \textbf{(4.15)}$$

Take note that the following properties are true:

$$\phi(t,t_0) = I \Leftrightarrow \phi^{-1}(t,s) = \phi(s,t)$$

Also, if $\phi(t)$ and $\psi(t)$ are fundamental matrix solutions of (4.1). There exists a non-singular constant C such that:

$$\phi(t) = \psi(t)C$$

## Adjoint System of Equations

A differential equation:

$$\dot{y}(t) = -y(t)A(t)$$

is called the adjoint system to (4.12), where $y(t)$ is a row n-vector [3,7]. If $\Phi(t)$ is a fundamental matrix of $\dot{x}(t) = A(t)x(t)$, then $\Phi^{-1}(t)$ is the fundamental matrix of the adjoint system $\dot{y}(t) = -y(t)A(t)$.

Proof

$$\Phi(t)\Phi^{-1}(t) = I$$

Differentiating with respect to $t$

$$\dot{\Phi}(t)\Phi^{-1}(t) + \Phi(t)\frac{d}{dt}\left[\Phi^{-1}(t)\right] = 0$$

$$\Rightarrow A(t)\Phi(t)\Phi^{-1}(t) + \Phi^{-1}(t) \bullet \frac{d}{dt}\left[\Phi^{-1}(t)\right] = 0$$

$$\Rightarrow \frac{d}{dt}\left[\Phi^{-1}(t)\right] = -\Phi^{-1}(t)A(t)$$

It implies that $\Phi^{-1}(t)$ satisfies $\dot{y}(t) = -y(t)A(t)$, it implies that $\Phi^{-1}(t)$ is a solution matrix of $\dot{y}(t) = -y(t)A(t)$. Next we have,

$$\frac{d}{dt}\left[\Phi(t)\Phi^{-1}(t)\right] = \det(I) = 1$$

It implies that

$$\det \Phi(t) \det \Phi^{-1}(t) = 1$$
$$\Rightarrow \det \Phi^{-1}(t) = [\det \Phi(t)]^{-1} \neq 0$$

*i.e.*

This completes the proof that $\Phi^{-1}(t)$ is a fundamental matrix of $\dot{y}(t) = -y(t)A(t)$. The general solution is $y = C\Phi^{-1}(t)$ .

## Remark 4.1

Consider the non-homogenous system:

$$\dot{y} = -yA + b(t)$$

The solution to the adjoint system satisfying $y(t_0) = y_0$ is given by:

$$y(t) = y_0 \Phi(t_0, t) + \int_{t_0}^{t} b(s)\Phi(t, s)ds$$

Proof

Set $y = C(t)\Phi^{-1}(t)$. Then

$$\dot{y} = \dot{C}(t)\Phi^{-1}(t) + C(t)\frac{d}{dt}\left[\Phi^{-1}(t)\right]$$
$$\Rightarrow -yA(t) + b(t) = \dot{C}(t)\Phi^{-1}(t) - C(t)\Phi^{-1}(t)A(t)$$

Therefore:

$$\dot{C}(t) = b(t)\Phi(t)$$
$$C(t) = C(t_0) + \int_{t_0}^{t} b(s)\Phi(s)ds$$

$y(t_0) = y_0 \Rightarrow C(t_0) = y_0\Phi(t_0)$ this implies that

$$y(t) = \left[ y_0 \Phi(t_0) + \int_{t_0}^{t} b(s)\Phi(s)ds \right] \Phi^{-1}(t)$$

$$= y_0 \Phi(t_0) + \Phi^{-1}(t) \int_{t_0}^{t} b(s)\Phi(s)ds$$

$y_0 \Phi(t,t_0) + \int_{t_0}^{t} b(s)\Phi(t,s)ds$ is the desired solution satisfying: $\Phi(s,t) = \Phi(s)\Phi^{-1}(t)$.

## Example 4.3

Consider the system:

$$\overset{\bullet}{x_1} = x_2$$

$$\overset{\bullet}{x_2} = -4x_1 - 5x_2$$

Find the fundamental matrix and adjoint equation of the system.

## Solution

$$A = \begin{pmatrix} 0 & 1 \\ -4 & -5 \end{pmatrix}$$

$$|A - I\lambda| = \begin{vmatrix} 0 - \lambda & 1 \\ -4 & -5 - \lambda \end{vmatrix} = (1 + \lambda)(4 + \lambda) = 0$$

The eigenvalue of companion matrix of $\lambda = -1$ and $\lambda = -4$. The eigenvectors associated with the eigenvalues are $\begin{pmatrix} 1 \\ -1 \end{pmatrix}$ and $\begin{pmatrix} 1 \\ 4 \end{pmatrix}$.

Let $\Phi(t)$ be a fundamental matrix of the system; then $\Phi(t) = [\phi_1(t), \phi_2(t)]$ where $\phi_1(t)$ and $\phi_2(t)$ are linearly independent solutions of the system.

Then:

$$\phi_1 = \begin{pmatrix} 1 \\ -1 \end{pmatrix} e^t = \begin{pmatrix} e^t \\ -e^t \end{pmatrix}, \phi_2 = \begin{pmatrix} 1 \\ -4 \end{pmatrix} e^{-4t} = \begin{pmatrix} e^{-4t} \\ -4e^t \end{pmatrix}. \Phi(t) = \begin{vmatrix} e^t & e^{-4t} \\ -e^t & -4e^{-4t} \end{vmatrix}$$

is the required fundamental matrix.

**Fundamental Matrix**

Let $y$ be a row vector, $y = [y_1, y_2]$, the adjoint equation for the system is:

$$\overset{\bullet}{y} = \begin{bmatrix} y_1, & y_2 \end{bmatrix} \begin{vmatrix} 0 & 1 \\ -4 & -5 \end{vmatrix} = \begin{bmatrix} 4y_2, & -y, +5y_2 \end{bmatrix}$$

*i.e.*

$$[\overset{\bullet}{y}_1, \overset{\bullet}{y}_2] = [4y_1, -5y_1 + 5y_2]$$

$$\overset{\bullet}{y}_1 = 4y_1$$

$$\overset{\bullet}{y}_2 = -y_1 + 5y_2$$

*i.e.* $\overset{\bullet}{y} = -yA$ and the desired fundamental matrix of the system is $\Phi^{-1}(t)$

$$\Phi^{-1}(t) = \frac{1}{-3e^{-2t}} \begin{pmatrix} \dfrac{4}{3} e^t & \dfrac{1}{3} e^t \\ e^{-t} & e^{-t} \end{pmatrix}$$

$$= \begin{pmatrix} \dfrac{4}{3} e^t & \dfrac{1}{3} e^t \\ -\dfrac{1}{3} e^{4t} & -\dfrac{1}{3} e^{4t} \end{pmatrix}$$

Meanwhile, Lemma 4.1 is instrumental to the study of the existence and uniqueness of solutions for initial value problems (IVP).

Existence and uniqueness theory is the cornerstone of differential equations. Therefore, particular attention will always be focused on it.

We consider matrix differential equations and how to use maple to solve them.

> *with*(*LinearAlgebra*);

> *with*(*ODETools*) :

> *with*(*plots*) :

> *A* := *Matrix*([[0, 1], [4, 5]]);

$$A := \begin{bmatrix} 0 & 1 \\ 4 & 5 \end{bmatrix}$$

> *CharacteristicPolynomial*(*A*, lambda);

$$\lambda^2 - 5\lambda - 4$$

> *solve*(%, lambda);

$$\frac{5}{2} + \frac{1}{2}\sqrt{41}, \frac{5}{2} - \frac{1}{2}\sqrt{41}$$

> *Eigenvalues*(*A*, *output* = 'list')

$$\left[ \frac{5}{2} + \frac{1}{2}\sqrt{41}, \frac{5}{2} - \frac{1}{2}\sqrt{41} \right]$$

> *Eigenvectors*(*A*)

$$\begin{bmatrix} \frac{5}{2} + \frac{1}{2}\sqrt{41} \\ \frac{5}{2} - \frac{1}{2}\sqrt{41} \end{bmatrix}, \begin{bmatrix} \dfrac{1}{\frac{5}{2} + \frac{1}{2}\sqrt{41}} & \dfrac{1}{\frac{5}{2} - \frac{1}{2}\sqrt{41}} \\ 1 & 1 \end{bmatrix}$$

> *B* := *Matrix*([[0, 1], [-4, -5]]);

$$B := \begin{bmatrix} 0 & 1 \\ -4 & -5 \end{bmatrix}$$

> $poly2 := CharacteristicPolynomial(B, \text{lambda});$

$$poly2 := \lambda^2 + 5\lambda + 4$$

> $Eigenvalues(B, output = 'list')$

$$[-1, -4]$$

> $Eigenvectors(B);$

$$\begin{bmatrix} -4 \\ -1 \end{bmatrix}, \begin{bmatrix} -\dfrac{1}{4} & -1 \\ 1 & 1 \end{bmatrix}$$

> $FundMatsol\_B := Matrix\left(\left[\left[-\dfrac{1e^{-4 \cdot t}}{4}, e^{-4 \cdot t}\right], [e^{-t}, e^{-t}]\right]\right);$

$$FundMatsol\_B := \begin{bmatrix} -\dfrac{1}{4}e^{-4t} & e^{-4t} \\ e^{-t} & e^{-t} \end{bmatrix}$$

> $C := Vector\left(\left[\exp\left(\dfrac{5}{2} + \dfrac{1}{2}\sqrt{41}\right) \cdot t, \exp\left(\dfrac{5}{2} + \dfrac{1}{2}\sqrt{41}\right) \cdot t\right]\right);$

$$C := \begin{bmatrix} e^{\frac{5}{2} + \frac{1}{2}\sqrt{41}} \, t \\ e^{\frac{5}{2} + \frac{1}{2}\sqrt{41}} \, t \end{bmatrix}$$

> $fundMatsol\_A := Matrix\left(\left[\left[\dfrac{1}{\dfrac{5}{2} + \dfrac{1}{2}\sqrt{41}}, \dfrac{1}{\dfrac{5}{2} - \dfrac{1}{2}\sqrt{41}}\right], [1, 1]\right]\right).C;$

$$fundMatsol\_A := \begin{vmatrix} \dfrac{e^{\frac{5}{2}+\frac{1}{2}\sqrt{41}}}{\frac{5}{2}+\frac{1}{2}\sqrt{41}}t + \dfrac{e^{\frac{5}{2}+\frac{1}{2}\sqrt{41}}}{\frac{5}{2}-\frac{1}{2}\sqrt{41}}t \\ 2e^{\frac{5}{2}+\frac{1}{2}\sqrt{41}}t \end{vmatrix}$$

## Lemma 4.2

$x(t)$ is the solution of differential equation (10.1), satisfying the initial condition:

$x(t_0) = x_0$ if and only if $x(t) = x_0 + \int_{t_0}^{t} f(s, x(s))ds, t \geq t_0, x(t_0) = x_0$.

## Proof

Suppose $x(t)$ is a solution of equation (4.1), then it must satisfy the differential equation by definition.

Thus:

$$\frac{dx(t)}{dt} = \frac{d}{dt} \int_{to}^{t} f(s, x(s))ds = f(t, x(t)).$$

Fundamental theorem of calculus+( see Lemma 4.3). Conversely if: $\dfrac{dx}{dt} = f(t, x(t))$
, integrating both sides yields:

$$\int \frac{dx(s)}{ds} ds = \int_{to}^{t} f(s, x(s))ds, t \geq t_o$$

$x(t) = x_o + \int_{to}^{t} f(s, x(x))ds$ which obeys equation (8.1) and so we are done.

## Lemma 4.3 (Fundamental Theorem of Calculus)

Let $f(t,x(t))$ defined in equation (4.1) be continuous on the interval $x_0 < x < b$. Given a number $x_0$, there exists only one solution, $x(t)$ of the equation (4.1) such that:

$$x(t) = x_o + \int_{to}^{t} f(s,x(s))ds, \ x_o = x(t_o)$$

## Normal Systems

Consider the general normal system of the first order differential equation:

$$\dot{x}_i = f_i(t,x),$$

Where $x$ is a vector, $x = (x_1, x_2, ..., x_n) \in E^n, i = 1, 2, ..., n$.

We are concerned in this section with the formulation and proof of a general theorem on the existence, uniqueness, continuality, and differentiability of solutions of the normal system (4.2).

The differential equation (1.1) is said to be autonomous or time-invariant. If it does not depend on time explicitly. Otherwise, it is said to be non-autonomous. If we set $t = x_{n+1}$ in (1.2), we obtain.

$$\dot{x}_i = f_i(x), x = (x_1, x_2, ..., x_n)$$ a non-autonomous system. This simply suggests a well-known fact that non-autonomous differential equation can be converted into non-autonomous differential equations and vice-versa by appropriate transformations.

## Lipchitz Condition

The concept of the Lipchitz condition provides a sufficient condition for the uniqueness, and existence of vector-valued function solution $X(t,x)$ of a differential in the normal form.

## Definition 4.2

A vector-valued function $X(t,x)$ satisfies a Lipchitz condition in a region R of $t,x$ space in $E^{n+1}$ if and only if, for some Lipchitz constant L,

$$|X(t,x) - X(t,y)| \leq L|x-y|, (t,x), (t,y) \in E^{n+1}.$$

A reformulation of this definition is as follows: let $X(t,x)$ be continuous on a compact set $U$ of D such that there exists a constant L for which the relation (4.1) is satisfied.

This definition in simplelanguage is equivalent to $f$ being lipchitzian $f(t,x)$ continuous together with the first partial derivative with respect to $x$ in the domain D and the Jacobian matrix:

$$\frac{\partial X(t,x)}{\partial dx_j}, j = 0,1,2,..n \text{ is bounded.}$$

This is a consequence of the mean value theorem.

Note that the Jacobian is defined as:

$$\frac{\partial X(t,x)}{\partial x_j} = \begin{vmatrix} \dfrac{\partial X_1}{\partial x_1} & \dfrac{\partial X_2}{\partial x_2} & \cdots & \dfrac{\partial X_n}{\partial x_n} \\ \dfrac{\partial X_1}{\partial x_2} & \dfrac{\partial X_2}{\partial x_2} & \cdots & \dfrac{\partial X_n}{\partial x_n} \\ \vdots & \vdots & \vdots & \vdots \\ \dfrac{\partial X_1}{\partial x_n} & \dfrac{\partial X_2}{\partial x_n} & \cdots & \dfrac{\partial X_n}{\partial x_n} \end{vmatrix}, X(t,x) = \left( X_1(t,x), X_2(t,x),..., X_n(t,x) \right)^T$$

We recall that the solutions of differential equations form a finite-dimensional vector space, $V(F)$ over a field $F$. Also, note also that every finitely generated space has a basis. Hence, there exists a solution basis for a solution space [1,3,7].

## Theorem 4.1

Let:

$$\varsigma(x) = a_n x^n + a_{n-1} x^{n-1} + \cdots + a_1 x + a_0 = 0 . \tag{4.17}$$

Then the solutions of equation (4.3) constitute a finitely generated vector space of dimension $n$.

## Proof

Let $W = \{x \in E^n : Lx = 0\}$, let $\alpha, \beta \in F, x, y \in W$.

Then:

$$\begin{aligned}
L(\alpha x + \beta y) &= \varsigma(\alpha x + \beta y) \\
&= \alpha(a_n x^n + a_{n-1} x^{n-1} + \cdots + a_1 x + a_0) \\
&\quad + \beta(a_n y^n + a_{n-1} y^{n-1} + \cdots + a_1 y + a_0) \\
&= \alpha(0) + \beta(0) = 0 \\
&\Rightarrow \alpha x, \beta y \in W
\end{aligned}$$

## Example 4.4

Consider $n = 1, x(0) = 2$, the solution of the IVP being $x(t) = \dfrac{2}{1 - 2t}$ existing on the interval $-\infty < t < \frac{1}{2}$.

Taking:

$$A(\alpha, \beta) = \{(t, x) : |x - 2| \le k, |t| < \infty\} \quad M = \sup_{x \in A} |f(t, x)| = \sup_{x \in A} (x^2) = (2 + k)^2$$

$\alpha = min\left(k, \dfrac{b}{m}\right) = \frac{1}{8}$, Since the largest positive number of $\dfrac{k}{(2 + k)^2}$ is $\frac{1}{8}$.

Therefore, the largest interval of existence of solutions is $|t| \le \frac{1}{8}$. However, the solution of the given IVP exists in the largest interval of existence. Then the solution can be extended over $0 \le t < \infty$. Noticing that $\varphi(x) = e^{-x}$ in the context of the theorem:

$$\lim_{\beta\to\infty}\int_{o}^{B}\frac{dx}{\varphi(x)} = \lim_{\beta\to\infty}\int_{o}^{B}e^{x}\,dx = \lim_{\beta\to\infty}(e^{B}-1) = \infty$$ By the Theorem 10.6, the assertion is justified.

## Definition 4.3

A solution $y(t)$ existing on I is said to be a maximum (upper) solution of the IVP:

$\dot{x}(t) = f(t,x(t))$, $x(t_{o}) = x_{o}$, for $t \in I, x(t) \le y(t)$, for every $t$. If, on the other hand, $w(t) \le y(t)$ for some solution $w(t)$ of the IVP we refer to it as a minimum (lower) solution of the IVP. Whenever an extreme (maximum or minimum) solution exists, it is always uniquely determined.

## Equi-Continuous Family

A family $\{\Phi_{\alpha}\}_{\alpha\in I}$ of functions in a Banach space $X$, I being index set, is point-wise bounded by a function $F(x)$, if $\|\Phi_{\alpha}\| \le F(x)$, for each $\alpha \in I$ for some , $x \in A$, a subset of X.

If there exists a common constant bound for the family $\{\Phi_{\alpha}\}_{\alpha\in I}$, we say it is uniformly bounded. In other words, if we can find a positive constant, $m$ such that $\|\Phi_{\alpha}\| \le m$ for every $\alpha \in I$, then the family $\{\Phi_{\alpha}\}_{\alpha\in I}$ is said to be uniformly bounded.

The family $\{\Phi_{\alpha}\}_{\alpha\in I}$ is equi-continuous if given $\in > 0$, there is a $\delta > 0$ such that:

$$0 < |x - y| < \delta \Rightarrow \|\Phi_{n}(x) - \Phi_{m}(y)\| < \in.$$

The choice of $\delta$ is independent of $x, y$ and $\Phi_{n}$. The Ascolis- Arzela's Theorem guarantees that every bounded equi-continuous family $\{\Phi_{\alpha}\}_{\alpha\in I}$ of functions in a compact Banach space has a uniformly convergent subsequence $\{\Phi_{nj}\}$. This inference will be exploited in the proof of the Carathedory Theorem. For further insight into Banach spaces and applications (See [6,8,9-11]).

## Carathedory Theorem

Let $f : \Omega \to E^n$ , where $\Omega$ is open in $E^n$. Suppose $f$ is dominated by a Lesbeque integrable function $m(t)$ on a compact subset of $\Omega$.

i.e.                                    $$\|f(t,x(t))\| \leq m(t)$$

Then, there exists a solution $x(t,t_0,x_0) \in \Omega$ of:

$$\dot{x}(t) = f(t,x(t)) \text{ passing through } (t_0,x_0), x(t_0) = x_0 \ .$$

## Proof

Let, $M(t) = \int\limits_{t_0}^{t} m(s)ds$ , clearly $0 \leq M(t)$ is a continuous non-decreasing function on $[t_0,t]$ . Define:

$$R(\alpha,\beta) = \left\{ (t,x) : |t - t_0| \leq a, \|x - x_0\| \beta \right\} \subset E^n ,$$

a compact subset of a Banach Space in $E^n$. Let $\{\phi_n\}_{n \geq 1}$ be a sequence of solutions to the above IVP.

Then:

$$\|\phi_n(t_2) - \phi_n(t_1)\| = \left\| \int\limits_{t_0}^{t_2} f(s,\phi_n(s))ds - \int\limits_{t_0}^{t_1} f(s,\phi_n(s))ds \right\|$$

$$\leq |M(t_2) - M(t_1)|$$

Since $M(t)$ is continuous, given $\varepsilon > 0, \exists \delta > 0$ such that:

$$0 < |t_2 - t_1| < \delta \Rightarrow |M(t_2) - M(t_1)| < \varepsilon$$

$$\therefore \|\phi_n(t_2) - \phi_n(t_1)\| \leq \varepsilon$$

Hence $\{\phi_n\}$ is an equi-continuous family in $C(\Omega, E^n)$.

$$\|\phi_n(t)\| \leq \|\phi_0\| + \left\|\int_{t_0}^{t} f(s, x(s))ds\right\|$$

$$\leq \|\phi_0\| + M(t) \leq (1+\beta)\|\phi_0\|$$

$$|M(t)| \leq \beta.$$

Therefore, $\|\phi_n(t)\| \leq \bar{M}, \bar{M}$ being a constant, $\|\phi_n\| < \infty, \{\phi_n\}$ is a uniformly bounded family. Ascolis- Arzela's theorem implies that there exists $\{\phi_{nj}\} \subset \{\phi_n\}$ such that $\phi_{nj} \to \phi$ uniformly.

$$\phi_{nj} = \phi_0 + \int_{t_0}^{t} f(s, \phi_{nj}(s))ds$$

By Lebesque dominated convergence theorem:

$$\int_{t_0}^{t} f(s, \phi_n(s))ds \to \int_{t_0}^{t} f(s, \phi(s))ds \text{ as } n \to \infty$$

Then:

$$\phi_{nj} = \phi_0 + \int_{t_0}^{t} f(s, \phi_{nj}(s))ds$$

$$\to \phi_0 + \int_{t_0}^{t} f(s, \phi(s))ds \text{ as } n \to \infty$$

Therefore,      $\phi(t) = \phi_0 + \int_{t_0}^{t} f(s, \phi(s))ds \text{ as } n \to \infty.$

## Example 4.5

Let $\qquad\qquad\qquad x \in C'([0,+\infty), E^1)$ and let:

$$\dot{x} = 3x + x\sin x, x(0) = x_0.$$

Show that:

$y(t) = |x_0|e^{-2t}$ is an upper bound for the solution of the above IVP and that $w(t) = |x_0|e^{-4t}$ is a lower bound for the solution.

Hence, show that $|x_0|e^{-4t} \le |x(t, t_0, x_0)| \le |x_0|e^{-2t}$.

## Solution

$$- \le \sin x \le 1 \text{ for every } x \in E^1 = (-\infty, +\infty)$$

$$\dot{x}(t) = -3x(t) + x(t)\sin x(t) \le -3x(t) + x(t) = -2x(t), x(t) > 0$$
$$\Rightarrow x(t) \le e^{-2t}|x_0| = y(t)$$

Similarly:

$$\dot{x}(t) = -3x(t) + x(t)\sin x(t) \ge -3x(t) - x(t) = -4x(t), x(t) < 0$$
$$\Rightarrow x(t) \ge e^{-4t}|x_0| = w(t)$$

$w(t) \le y(t)$. Hence the proof.

## Problem 4.1

1. If $x \in C'([0,+\infty), E^1)$ and let:

$$\dot{x} = 4x + x\cos x, x(0) = x_0$$

Show that $0 \le |x_0| e^{-5t} \le |x(t)| \le |x_0| e^{-3t}$ for all $t \ge 0$

2.  Let $x \in C^1([0,+\infty), E^1)$ and let:

$$\dot{x} = x^2 - t \text{ for } t \ge 0, |x| \ge 1, x(0) = 1.$$

Show that $x = \dfrac{1}{1-t}, t > 1$

is the upper bound for the solution and $x_l = 1+t$ is a lower bound for the solution of the differential equation.

3. Consider the IVP:

$$\dot{x}(t) = f(t, x(t)), t > 0, x(0) = x_0. \text{ Is the function:}$$

$$\phi(t) = \begin{cases} 0 & t < 4 \\ (t-4)^2 & t \ge 4 \end{cases}$$

an upper or a lower solution of $\dot{x}(t) = 2\sqrt{x(t)}, x(0) = 0$? Justify your claim.

**Problem 4.2**

1. Investigate whether the following are Lipchitzian with respect to $x$:

a.                                    $f(t,x) = tx^2$

b.                                    $f(t,x) = t \sin x$

c.                                    $f(t,x) = \dfrac{1}{\sqrt{2}}$

d.                                    $f(t,x) = t^2 \cos x$

If so, find the Lipchitz constant.

2. What can you say about the existence and uniqueness of the solutions of the IVPs:

a.
$$\dot{x}(t) = \frac{1}{\sqrt{x}}, x(0) = -1$$

b.
$$\dot{x}(t) = tx^2, x(0) = x_0$$

c.
$$\dot{x}(t) = t\sin x, x(0) = 1$$

d.
$$\dot{x}(t) = t\cos x, x(0) = 1$$

e.
$$\dot{x}(t) = t^2\sqrt{x}, x(0) = 1$$

3. a. State the theorem on uniqueness of solution of the IVP: $\dot{y}(t) = F(t, y(t)), y(t_0) = y_0$

b. Check whether the function:

$$F(y_1, y_2) = \begin{bmatrix} \sqrt{1 + y_2^2} \\ \sqrt{1 + y_2^2} - y_2 \end{bmatrix}, (y_1, y_2) \varepsilon E^2$$

fulfills the assumptions of the above theorem.

## CONCLUSION

Here, in this chapter, we treated topics like, norm of matrices and how to obtain a principal matrix, and exponential matrix solutions to autonomous and non-autonomous differential equations. Maple software is demonstrated to construct a fundamental matrix solution to some vector differential equations. Lipschitz condition, equi-continuous family of any existence and Caratheodory theorem are used to obtain the solution of ODES.

# REFERENCES

[1]   W.E. Boyce, and R.C. Diprima, *Elementary Differential Equations,* 7th ed John Wiley and Son, 1977.

[2]   H. Brain Chirgwin, and Plumpton. Charles, *Advanced Theoretical Mechanics: A Course of Mathematics.* Pergamon Press Ltd , 1961.

[3]   A. Coddington Earl, and Levinson Norman, *Theory of Ordinary Differential Equations.* McGraw Hill, 1955.

[4]   Kreyszig Erwin, *Advanced Engineering Mathematics.* John Wiley Publication: USA, 2000.

[5]   Birkhoff Garrett, and Rota Gian-Carlo, *Ordinary Differential Equations.* Wiley, 1989.

[6]   V.C.L. Huston, and J.J. Pym, *Application of Functional Analysis and Operator theory.* Academic Press London: New-York / Toronto, 1989.

[7]   K. Hale Jack, *Ordinary Differential Equations.* Wiley-Interscience: New York, London, Sydney, Toronto, 1969.

[8]   E. Kreyszig, *Introduction to Functional Analysis with Application.* Wiley Indian publication, 2007.

[9]   V. Lakshmikantam, D.D. Bainov, and P.S. Simeonov, *Theory of Impulsive Differential Equations.* World Scientific Publications: Singapore, 1989.

[10]  B.O. Oyelami, S.O. Ale, and M.S. Sesay, "On existence", *J. Niger. Math. Soc.,* vol. 21, pp. 13-25, 2002.

[11]  B.O. Oyelami, and S.O. Ale, "On existence of solution, oscillation and non-oscillation properties of delay equations containing 'Maximum'", In: *Acta Appl. Math. J.,* vol. 109. 2010, pp. 683-701.

# CHAPTER 5

# Canonical Transformation and Matrix Solutions of Differential Equations

**Abstract:** We consider canonical transformation for transforming scalar differential equations to matrix differential equations. We determine conditions for linear independence of solutions using the Wronskian method and use the Jordan canonical form to find bounds for solutions of ODES. Also considered are: the generalized eigenvectors method for obtaining matrix solutions to ODES and corresponding bounds for the autonomous differential equations, upper and lower bounds for solutions. Conditions for continuous dependence of solutions on initial data are formulated. Periodic systems are studied too with the application of the Floquet rule to finding solutions to some linear periodic systems. The Theorem on how to construct monodromy matrices is presented for the linear periodic systems together with some examples.

**Keywords:** Autonomous differential equations, Canonical transformation, Floquet rule, Jordan canonical forms, Matrix solutions, Monodromy matrices, ODES solutions, Periodic systems, Upper and lower bounds, Wronskian method.

## INTRODUCTION

This chapter looks at the canonical transformation method, and transformations of scalar equations into vector forms [1-5]. The fundamental matrix, principal matrix, and adjoint to homogenous systems are revisited. The theory of autonomous linear homogeneous systems will be introduced together with the canonical transform process in Jordan Canonical form. We will construct fundamental matrix solutions using Sylvester's formula and derive upper and lower solution bounds to ODES. We will investigate how solutions of ODEs continuously depend on initial data. At the end of the chapter, linear periodic systems and applications are to be considered with examples given.

## Canonical Transformation

Suppose we have a scalar differential equation:

$$D^n y + a_1(t)D^{n-1} + ... + a_n(t)y = g(t) \tag{5.1}$$

**Benjamin Oyediran Oyelami**
**All rights reserved-© 2024 Bentham Science Publishers**

Our interest is to find an equivalent vector equation having the same solution as the equation (5.1).

Canonical transformation provides us with a methodology.

Let:

$$
\left.
\begin{aligned}
& y = y_1 \\
& Dy = Dy_1 = y_2 \\
& D^2y = D^2y_1 = Dy_2 = y_3 \\
& \vdots \\
& \text{Then} \\
& D^{n-1}y = -a_ny_1 - a_{n-1}y_2 - \cdots - a_1y_2 + g(t)
\end{aligned}
\right\}
\tag{5.2}
$$

Let $x = (y_1, y_2, ..., y_n)^T$ then:

$$
\dot{x} = A(t)x + h(t)
\tag{5.3}
$$

Where:

$$
A(t) = \left[ a_{ij} \right]_{i,j \in \{1,2,...,n\}} =
\begin{bmatrix}
0 & 1 & 0 & \cdots & 0 & 0 \\
0 & 0 & 1 & 0 & \cdots & 0 \\
0 & 0 & 0 & 1 & \cdots & 0 \\
\vdots & \vdots & \vdots & \vdots & \cdots & \vdots \\
0 & 0 & 0 & 0 & \cdots & 1 \\
-a_n & -a_{n-1} & a_{n-2} & \cdots & -a_2 & -a_1
\end{bmatrix}
\tag{5.4}
$$

$h(t) = [0, 0, \cdots, g(t)]^T$

We recall the definition of Wronskian of $[\phi_1, \phi_2, \cdots, \phi_n]$ of (n) continuously differentiable functions:

$$W[\phi_1,\phi_2,\cdots,\phi_n] = \begin{vmatrix} \phi_1 & \phi_2 & \cdots & \phi_n \\ \dot{\phi_1} & \dot{\phi_2} & \cdots & \dot{\phi_n} \\ \vdots & \vdots & \vdots & \vdots \\ \phi_1^{(n-1)} & \phi_2^{(n-1)} & \cdots & \phi_n^{(n-1)} \end{vmatrix}$$ (5.5)

We said in (Chapter 3 ,section 3) that $\phi_1,\phi_2,\cdots,\phi_n$ are linearly independent on the interval $[t_0,+\infty)$ if $W[\phi_1,\phi_2,\cdots,\phi_n]$ does not vanish for at least one $t \in [t_0,+\infty)$. Otherwise, triviality of the Wronskian $W[\phi_1,\phi_2,\cdots,\phi_n]$ implies the linear dependence of $[\phi_1,\phi_2,\cdots,\phi_n]$ (see [6, 8]).

**Theorem 5.1**

Let $\dot{x}(t) = A(t)x(t), x(t_0) = x_0$, $t \in [t_0,+\infty)$, where $A(t)$ is an n-square matrix function of $t$ on the interval $J = [0,+\infty)$. By close analogy, the exponential matrix is similar to the scalar exponential function and shared same properties with the exception of commutativity which breaks down in the case of exponential matrices.

Let $A, B$ be $n \times n$ matrices. Then the following are true:

$$1.\, e^{A+B} = e^A \cdot e^B$$ (5.6)

$$2.\, (e^A)^{-1} = e^{-A} \quad \text{(inverse)}$$ (5.7)

**Warning**

No mistake should be committed in assuming commutativity for $e^A$ and $e^B$. In general, it is not guaranteed. In 1927, N H Abel established a relation for obtaining the Wronskian of solutions of differential equations using the traces of matrices of second order equations. The ideal was later generalized by J. Liouville and M. V. Ostrogradsky to nth order equations [7].

**Theorem 5.1**

If $x_1, x_2, ..., x_n$ are solutions of equation (5.3),

then $W[x_1, x_2, \cdots, x_n] = W[x_1, x_2, \cdots, x_n](a) \exp - \int_0^t a_1(s) ds$. It is quite simple to observe from (5.5) that the solutions $x_1, x_2, ..., x_n$ of (5.3) are linearly independent if $W[x_1, x_2, \cdots, x_n](a) \neq 0$.

## Proof

We assume the reader is familiar with the differentiation of determinant. The above definition can be applied to the derivation of determinants of the fundamental matrices of homogenous linear systems. In this regard, suppose $X(t)$ is a fundamental matrix of $\dot{x} = A(t)x, t \in [0, +\infty)$.

Then, in particular $\dot{X}(t) = A(t)X(t), t \in [0, +\infty)$ and $\det X(t) \equiv 0, \forall t \in [0, +\infty)$. Set $X(t) = (x_1(t), x_2(t), ..., x_n(t))$, where $\dot{x}_j = A(t)x_j, j \in \{1, 2, \cdots, n\}$

and $x_j = \begin{pmatrix} x_{1j} \\ x_{2j} \\ \vdots \\ x_{nj} \end{pmatrix}, j \in \{1, 2, \cdots, n\}$. Let $A(t) = \left( a_{ij} \right)_{i,j=1,2,3,\cdots n}$ Then:

$$\dot{x}_j = A(t)x_j \Rightarrow \dot{x}_j = \sum_{i=1}^n a_{ij} x_j .$$

$$\det X(t) = \det \begin{bmatrix} x_{11} & x_{12} & \cdots & x_{1n} \\ x_{21} & x_{22} & \cdots & x_{2n} \\ \vdots & \vdots & \cdots & \vdots \\ x_{n1} & x_{n2} & \cdots & x_{nn} \end{bmatrix} = \begin{vmatrix} x_{11} & x_{12} & \cdots & x_{1n} \\ x_{21} & x_{22} & \cdots & x_{2n} \\ \vdots & \vdots & \cdots & \vdots \\ x_{n1} & x_{n2} & \cdots & x_{nn} \end{vmatrix}$$

$$\frac{d}{dt}\left[\det X(t)\right] = \begin{vmatrix} \dot{x}_{11} & \dot{x}_{12} & \cdots & \dot{x}_{1n} \\ x_{21} & x_{22} & \cdots & x_{2n} \\ \vdots & \vdots & \cdots & \vdots \\ x_{n1} & x_{n2} & \cdots & x_{nn} \end{vmatrix}$$

$$= \begin{vmatrix} x_{11} & x_{12} & \cdots & x_{1n} \\ \dot{x}_{21} & \dot{x}_{21} & \cdots & \dot{x}_{21} \\ \vdots & \vdots & \cdots & \vdots \\ x_{n1} & x_{n2} & \cdots & x_{nn} \end{vmatrix} + \cdots + \begin{vmatrix} x_{11} & x_{12} & \cdots & x_{1n} \\ x_{21} & x_{21} & \cdots & x_{21} \\ \vdots & \vdots & \cdots & \vdots \\ \dot{x}_{n1} & \dot{x}_{n2} & \cdots & \dot{x}_{nn} \end{vmatrix} \tag{5.8}$$

$$= D_1 + D_2 + \cdots + D_n$$

Consider the case when $n = 3$ yields:

$$\frac{d}{dt}\left[\det X(t)\right] = \begin{vmatrix} \dot{x}_{11} & \dot{x}_{12} & \dot{x}_{13} \\ x_{21} & x_{22} & x_{23} \\ x_{31} & x_{32} & x_{33} \end{vmatrix} + \begin{vmatrix} x_{11} & x_{12} & x_{13} \\ \dot{x}_{21} & \dot{x}_{22} & \dot{x}_{23} \\ x_{31} & x_{32} & x_{33} \end{vmatrix} + \begin{vmatrix} x_{11} & x_{12} & x_{13} \\ x_{21} & x_{22} & x_{23} \\ \dot{x}_{31} & \dot{x}_{32} & \dot{x}_{33} \end{vmatrix}$$

$$= D_1 + D_2 + D_3$$

Clearly:

$$D_1 = \begin{vmatrix} \sum_{i=1}^{3} a_{ik}x_{k1} & \sum_{i=1}^{3} a_{ik}x_{k2} & \sum_{i=1}^{3} a_{ik}x_{k3} \\ x_{21} & x_{22} & x_{23} \\ x_{31} & x_{32} & x_{13} \end{vmatrix}$$

$$= \begin{vmatrix} a_{11}x_{11} + a_{12}x_{21} + a_{13}x_{31} & a_{12}x_{21} + a_{12}x_{22} + a_{13}x_{33} & a_{11}x_{13} + a_{12}x_{23} + a_{13}x_{33} \\ x_{21} & x_{22} & x_{23} \\ x_{31} & x_{32} & x_{13} \end{vmatrix}$$

Therefore:

$$\frac{d}{dt}X(t) = [a_{11} + a_{22} + \cdots + a_{nn}]\det X(t) \quad \frac{d}{dt}X(t) = [a_{11} + a_{22} + \ldots + a_{nn}]\det X(t) \quad \textbf{(5.9)}$$
$$\mathrm{Tr}A(t)\det X(t)$$

Integration gives:

$$\det X(t) = \det X(a)\exp\left(\int_a^t \mathrm{Tr}A(s)ds\right) \quad \textbf{(5.10)}$$

$\mathrm{Tr}A(t) = \left[a_{11} + a_{22} + \cdots + a_{nn}\right]$, the sum of the diagonal entries of matrix $A(t)$.

**Autonomous Linear Homogenous System**

An autonomous linear homogenous system is as follows:

$$\dot{x} = Ax , \quad \textbf{(5.11)}$$

where in the case, $f(x) = A(t)x, A$ is constant $n \times n$ matrix. The target sum of this section is to obtain an estimation for the fundamental matrix $e^{At}$ of (5.11).

This necessarily motivates preliminary discussion on Sylvester's and Jordan canonical forms of representation of $e^{At}$, accompanied by relevant theorem.

**Sylvester's Formula**

If the eigenvalues of A are all distinct, then the fundamental matrix $X(t) = e^{A(t-t_0)}$ of the equation (5.11) such that $x(t_0) = I$ is given by:

$$e^{A(t-t_0)} = \sum_{i=1}^n a_i p_i(\lambda)e^{\lambda_i(t-t_0)} \quad \textbf{(5.12)}$$

Where $p_i(\lambda) = \prod\limits_{j=1, j\neq i}^{n}\left(\dfrac{\lambda - \lambda_j}{\lambda_i - \lambda_j}\right), \prod\limits_{i=j}^{n}$ is the product notation [See Jordon [28],49 – 51]

for the proof if:

$$A = diag(d_1, d_2, \cdots, d_n) = \begin{bmatrix} d_1 & 0 & \cdots & 0 \\ 0 & d_2 & \cdots & 0 \\ \vdots & \vdots & \ddots & \vdots \\ 0 & 0 & \cdots & d_n \end{bmatrix}$$

Then $e^{A(t-t_0)=diag(e^{d_1(t-t_0)},e^{d_2(t-t_0)},\cdots,e^{d_n(t-t_0)})}$ whether or not $\lambda_i$ is distinct.

### Example 5.1

Find $e^{At}$ for the system:

$$\dot{x} = 2x - y$$
$$\dot{y} = 2x + y$$

### Solution

In vector form, the system is equivalent to:

$$\dot{z} = Az$$

$$A = \begin{pmatrix} 2 & -1 \\ -2 & 3 \end{pmatrix}, \quad z = \begin{pmatrix} x \\ y \end{pmatrix}$$

The characteristic equation is:

$$0 = |A - \lambda I| = \begin{vmatrix} 2-\lambda & -1 \\ -2 & 3-\lambda \end{vmatrix} = (\lambda - 1)(\lambda - 4)$$

Therefore, $\lambda_1 = 1, \lambda_2 = 4$. Using the Sylvester's formula, we have:

$$e^{At} = \sum_{i=1}^{2} a_i p_i(A) e^{\lambda_i t}$$

where $a_1 \, p_1(A)$ and $a_2 \, p_2(A)$ can be computed as follows :

$$a_1 p_1(A) = \frac{A - \lambda_2 I}{\lambda_1 - \lambda_2} = -\frac{1}{3} \begin{pmatrix} -2 & -1 \\ -2 & -1 \end{pmatrix}$$

$$a_2 p_2(A) = \frac{A - \lambda_1 I}{\lambda_2 - \lambda_1} = \frac{1}{3} \begin{pmatrix} 1 & -1 \\ -2 & 2 \end{pmatrix}$$

$$e^{At} = \frac{1}{3} \begin{pmatrix} 1 & -1 \\ -2 & 2 \end{pmatrix} e^{At} - \frac{1}{3} \begin{pmatrix} -2 & -1 \\ -2 & -1 \end{pmatrix} e^{t}$$

$$= \frac{1}{3} \begin{pmatrix} 2e^t + e^t & \left(e^t - e^{4t}\right) \\ 2\left(e^t - e^{4t}\right) & \left(e^t + 2e^{4t}\right) \end{pmatrix}$$

## Remark 5.1

$e^{At}$ could as well be obtained by diagonaling the companion matrix by the use of Jordan Canonical form.

## Jordan Canonical Form

For any $n \times n$ real matrix, there exists a non-singular matrix, P such that $P^{-1}AP = J$ or $A = PJP^{-1}$, where J is in the canonical form [6]:

$$J = \begin{pmatrix} J_0 & & 0 \\ & J_1 & \\ 0 & & J_s \end{pmatrix} = diag\left(J_0, J_1, J_s\right)$$

and
$$J_0 = \begin{pmatrix} \lambda_1 & & & 0 \\ & \lambda_2 & & \\ & & \lambda_2 & \\ 0 & & & \lambda_s \end{pmatrix}$$

$\lambda_1, \lambda_2, ..., \lambda_k$ are necessarily distinct.

$$J_i = \begin{pmatrix} \lambda_{k+i} & 0 & 0 & .. & 0 \\ 0 & \lambda_{k+i} & & & \\ 0 & & \lambda_{k+i} & & \\ 0 & 0 & & .... & \lambda_{k+i} \end{pmatrix}$$

$J_i$ is the maximum matrix for each $i$ , $\lambda_{k+i}$ needs not be different from $\lambda_{k+j}$ *for* $i \neq j$　$n_1 + n_2 + ......n_f = n$

$J_i$ is in Jordan Canonical form. Clearly:

$$J_i = \lambda_{k+i} I_{ni} + N_{ni}, \quad i = 1,2,....3$$

where $I_{ni}$ is the identity matrix of order $n_i$ and $N_{ni}$ is the $ni \times ni$ nilpotent matrix,

$$\begin{pmatrix} 0 & 1 & 0 & \cdots & 0 \\ 0 & 0 & 1 & \cdots & \\ \vdots & \vdots & \vdots & \cdots & 0 \\ 0 & 0 & 0 & \cdots & 1 \\ 0 & 0 & 0 & 0 & 0 \end{pmatrix}_{ni \times ni}$$

$$e^{Jit} = e^{\lambda_{K+l}t} \, e^{Nnit}, \, l+1,2,\ldots\ldots s.$$

$$e^{Nit} = \begin{pmatrix} 1 & t & t^2\!/2 & \ldots & 0 & 0 & \dfrac{t^{ni}-1}{(ni-1)!} \\ 0 & 1 & t & & & & \\ \vdots & & \cdot & & & & \dfrac{t^{n_1-2}}{(ni-2)!} \\ & & & \vdots & & & \\ 0 & 0 & 0 & & & & t \end{pmatrix}$$

$$N_l = \begin{pmatrix} 0 & 1 & 0 & \ldots & 0 & 0 \\ 0 & 0 & 1 & & 0 & 0 \\ 0 & 0 & 0 & & 0 & 1 \\ \vdots & & \vdots & & \vdots & \vdots \\ 0 & 0 & 0 & & 0 & 0 \end{pmatrix}, \; \textit{Nilpotent matrix}$$

[See Jack [8], pp 99]

Obtain $e^{At}$ for a system with $A = \begin{pmatrix} 2 & -1 \\ -2 & 3 \end{pmatrix}$ using the Jordan Canonical form.

**Solution**

We have $A = \begin{pmatrix} 2 & -1 \\ -2 & 3 \end{pmatrix}$ and eigenvalues are found to be $\lambda_1 = 1$, $\lambda_2 = 4$. The corresponding eigenvalues are $p_1 = col(1;1)$, $p_2 = (-1,2)$ *respectively* ,then:

$$e^{At} = p^{-1}e^{Jt}p$$

where $e^{Jt} = diag\left(e^t, e^{4t}\right)$, $p = (p_1, p_2)$.

Hence:

$$e^{At} = \frac{1}{3}\begin{pmatrix} 1 & -1 \\ 1 & 2 \end{pmatrix}\begin{pmatrix} e^t & 0 \\ 0 & e^{4t} \end{pmatrix}$$

$$= \frac{1}{3}\begin{bmatrix} 2e^t + e^{4t} & \left(e^t - e^{4t}\right) \\ 2e^t + e^{4t} & \left(e^t + 2e^{4t}\right) \end{bmatrix}$$

This is no doubt that the same representation is obtained in example using Sylvester's formula.

**Remark 5.2**

Each $J_i$ is called a Jordan block. $J_0$ is also a Jordan block and each such block contains an eigenvalue of A.

The following Lemma is relevant:

**Lemma 5.1**

If a matrix P exists such that:

$A = PJP^{-1}$ where $J$ is a diagonal

$A^K = PJ^K P^{-1}$ for any positive integer $k$ if $A$ is non-singular the result is true
        for any negative integer, ie., in particular

$A^{-1} = PJ^{-1}P^{-1}$

**Proof**

For any positive integer k.

$$A = PJP^{-1}$$
$$A^K = \left(PJP^{-1}\right)\left(PJP^{-1}\right).\left(PJP^{-1}\right)$$

$$A^K = PJP^{-1}PJP^{-1}...PJP^{-1}$$

A product of k equal times:

$$= PJ^K P^{-1}$$

For k= -1, $A^{-1} = (PJP^{-1})^{-1}$

$$=(P^{-1})^{-1}\, J^{-1}P^{-1}$$

$$A^{-1} = PJP^{-1}$$

$V_1, V_2, \cdots, V_n$ are $n \times n$ matrices, $P = (V_1, V_2, \cdots, V_n)$ which transforms $A$ into the J, Jordan Canonical.

From:

$$J = P^{-1}AP = \begin{pmatrix} J_0 & & 0 \\ : & J_1 & \\ 0 & & J_s \end{pmatrix}$$

Hence:

$$e^{Jt} = \begin{pmatrix} e^{J_0 t} & & 0 \\ & e^{J_1 t} & \\ 0 & & e^{J_s t} \end{pmatrix}$$

$$e^{J_0 t} = \begin{pmatrix} e^{\lambda_1 t} & & 0 \\ & e^{\lambda_2 t} & \\ 0 & & e^{\lambda_n t} \end{pmatrix}$$

## Theorem 5.2

If all real part of the eigenvalues of A are less than or equal to $\alpha$ and all eigenvalues are distinct, then there exists a real number $C \geq 1$. Such that:

$$\|e^{At}\| \leq ce^{\alpha t}, \forall t \in J \tag{5.13}$$

## Proof

(1)      If A is diagonal, $A = \begin{pmatrix} a_{11} & \cdots & 0 \\ \vdots & a_{22} & \cdots \\ 0 & & a_{nn} \end{pmatrix}$,

Then:

$$e^{At} = diag\left(e^{\alpha_1 t}, e^{\alpha_2 t}, \cdots, e^{\alpha_n t}\right)$$

and

$$\left\|e^{At}\right\| \leq \max\left|e^{a_{11} t}\right| = e^{t \max \ Re(a_{11})}$$
$$\leq e^{\alpha t}$$

Since                 $Re\left(a_{11}\right) \leq \alpha$. *Hence* $C = 1$.

(2) If A is not diagonal, then: $A = JPJ^l$ *where J is the Jordan Camonical form of A. In this case*

$$e^{At} = \sum_{k=0}^{\infty} \frac{(At)^k}{k!}$$
$$= P\left(\sum_{k=0}^{\infty} \frac{(Jt)^k}{k!}\right) P^{-1}$$
$$= Pe^{-Jt} P^{-1}$$
$$\left\|e^{At}\right\| \leq \left\|Pe^{-Jt} P^{-1}\right\| \leq \|P\| \left\|e^{-Jt}\right\| \left\|P^{-1}\right\| = c\left\|e^{-Jt}\right\|$$
$$c = \|P\| \left\|P^{-1}\right\| \geq 1$$

## Method of Generalized Eigenvectors

Finding exponential matrices in higher dimension is not always an easy task. The method of generalized eigenvectors gives a generalized technique for obtaining eigenvectors and generalized eigenvectors associated with eigenvalues

$\lambda_1, \lambda_2, \cdots, \lambda_s$ multiplicities $n_1, n_2, \cdots, n_s$ respectively.

A vector V is called a generalized eigenvector of order $k$ if:

$$(A-\lambda I)^k V = 0, (A-\lambda I)^{k-1} V \neq 0$$

**Method**

$$Let \; V_K = V$$

$$V_{K-1} = (A-\lambda I)V = (A-\lambda I)V_K$$

$$V_1 = (A-\lambda I)V = (A-\lambda I)V_2$$

The set $\{v_1, V_2, \cdots, V_n\}$ is the set of generalized eigenvectors.

The following are properties of generalized eigenvectors:

(i) $\{V_k\}, k=1,2,...,n$ are linearly independent. That is generalized eigenvector associated with distinct $\lambda_1$ are linearly independent.

(ii) Let U and V be generalized eigenvectors corresponding to the eigenvalue $\lambda_i$

If:
$$U_i = (A-\lambda I)^{k-i} U, \; i=1,2,......k$$
$$V_i = (A-\lambda I)^{i-j} V, \; j=1,...l$$

such that $U, V$ are linearly independent then $\{U\}, \{V\}$ are linearly independent $i \leq l \leq k, i, \; j \leq i...$

Algorithm ( step by step computation). The generalized eigenvector U of rank $k$ is given by $(A-\lambda I)^k U = 0$, where rank $(A-\lambda I)^k = $ rank $(A-\lambda I)^{k+1}$

That is, nullity $(A-\lambda I)^k = $ nullity $(A-\lambda I)^{k+1}$

Define:

$$U_i = (A-\lambda I)^{k-i} U, \; i=1,2,.....n$$

If $k = n$, find another linearly independent generalized eigenvector with the largest possible rank. Try to find another eigenvector of rank $k$; if this is not possible, try for rank $k - 1$ until $ni$ linearly independent generalized eigenvectors are determined.

**Assumption**

$\lambda_1, \lambda_2,.....\lambda_s$ are distinct eigenvalues with multiplicities $n_1, n_2, \cdots, n_s$ respectively. The computation of $n_i$ linearly independent generalized eigenvectors of A associated with $\lambda$ will now be undertaken.

**Note**

If rank $(A - \lambda I) = r$, there exist $n - r$ chains of generalized eigenvectors associated with $\lambda$; repeat the process for $\lambda_1, \lambda_2, \cdots, \lambda_s$.

**Example 5.2**

Obtain $e^{At}$ for the problem $\dot{x} = Ax$, where $x \in E^1$.

$$A = \begin{vmatrix} -2 & 0 & 0 & 0 & 0 & 0 \\ 0 & -1 & 1 & 0 & 0 & 0 \\ 0 & 0 & -1 & 0 & 0 & 0 \\ 0 & 0 & 0 & 3 & 0 & 0 \\ 0 & 0 & 0 & 0 & 3 & 1 \\ 0 & 0 & 0 & 0 & 0 & 3 \end{vmatrix}$$

**Solution**

$$\begin{vmatrix} -2 & 1 & & & & \\ & -1 & 1 & & & \\ & 0 & -1 & & & \\ & & & 3 & & \\ & & & & 3 & 1 \\ & & & & 0 & 3 \end{vmatrix} , \; J_0 = [-2], J_1 = \begin{bmatrix} -1 & 1 \\ 0 & -1 \end{bmatrix}$$

$$J_2 = [3], \quad J_3 = \begin{vmatrix} 3 & 1 \\ 0 & 3 \end{vmatrix}$$

$$e^{J_0 t} = e^{-2t}$$

$$J_1 = -1 \begin{bmatrix} 1 & 0 \\ 0 & 1 \end{bmatrix} + \begin{bmatrix} 0 & 1 \\ 0 & 0 \end{bmatrix}, \quad e^{J_1 t} = e^{-t} \begin{bmatrix} 1 & t \\ 0 & 1 \end{bmatrix}$$

$$J_1 = -1 \begin{vmatrix} 1 & 0 \\ 0 & 1 \end{vmatrix} + \begin{vmatrix} 1 & 0 \\ 0 & 0 \end{vmatrix}, e^{-t} \begin{vmatrix} 1 & t \\ 0 & 1 \end{vmatrix}$$

$$e^{J_2 t} = e^{3t}$$

$$e^{J_3 t} = e^{3t} \begin{bmatrix} 1 & t \\ 0 & 1 \end{bmatrix}$$

Nothing that $e^{N_i t} = e^{\lambda_i t} e^{N_i t}$, $e^{N_i t}$ is the exponent of a nilpotent matrix.

$$e^{Jt} = diag\left[ e^{J_0 t}, e^{J_1 t}, \cdots, e^{J_n t} \right], e^{At} = P^{-1} e^{Jt} P, P = I .$$

## Example 5.3

Obtain $e^{At}$ for the problem $\underline{\dot{x} = Ax}$ where $x \in E^6$

$$A = \begin{vmatrix} -2 & 0 & 0 & 0 & 0 & 0 \\ 0 & -1 & 1 & 0 & 0 & 0 \\ 0 & 0 & -1 & 0 & 0 & 0 \\ 0 & 0 & 0 & 3 & 0 & 0 \\ 0 & 0 & 0 & 0 & 3 & 1 \\ 0 & 0 & 0 & 0 & 0 & 3 \end{vmatrix}$$

## Solution

$$
\begin{vmatrix}
-2 & & & & \\
& 0 & -1 & & \\
& & 3 & & \\
& & & 3 & 1 \\
& & & 0 & 3
\end{vmatrix}
$$

$$
J_0 = \begin{bmatrix} -2 \end{bmatrix}, \; J_1 = \begin{bmatrix} -1 & 1 \\ 0 & -1 \end{bmatrix}, \; J_2 = \begin{bmatrix} 3 \end{bmatrix}, \; J_3 = \begin{bmatrix} 3 & 1 \\ 0 & 3 \end{bmatrix}
$$

$$
e^{J_0 t} = e^{-2t}
$$

$$
J_1 = -1 \begin{bmatrix} 1 & 0 \\ 0 & 1 \end{bmatrix} + \begin{bmatrix} 0 & 1 \\ 0 & 0 \end{bmatrix}, \; e^{Jt} = e^{-t} \begin{bmatrix} 1 & t \\ 0 & 1 \end{bmatrix}
$$

$$
e^{J_2 t} = e^{3t}
$$

$$
e^{J_3 t} = e^{3t} \begin{bmatrix} 1 & t \\ 0 & 1 \end{bmatrix}
$$

Note that $e^{J_1 t} = e^{\lambda_2 t} e^{N_2 t}$

$$
e^{Jt} = \begin{pmatrix}
e^{J_0 t} & & & \\
& e^{J_1 t} & & \\
& & e^{J_2 t} & \\
& & & e^{J_0 t}
\end{pmatrix}
= \begin{bmatrix}
e^{-2t} & & & & \\
& e^{-4t} & e^{-4t} & & \\
& 0 & e^{-4t} & & \\
& & & e^{3t} & \\
& & & e^{3t} & t^{e3} \\
& & & 0 & e^{3t}
\end{bmatrix}
$$

Using the fact that:

$$e^{At} = P^{-1} e^{Jt} P \quad but \quad P = I$$

$$e^{At} = e^{Jt} = \begin{bmatrix} e^{-2t} & 0 & 0 & 0 & 0 & 0 \\ 0 & e^{-t} & te^{-t} & 0 & 0 & 0 \\ 0 & 0 & e^{-t} & 0 & 0 & 0 \\ 0 & 0 & 0 & e^{3t} & 0 & 0 \\ 0 & 0 & 0 & 0 & e^{3t} & te^{3t} \\ 0 & 0 & 0 & 0 & 0 & e^{3t} \end{bmatrix}$$

But,

$$\lambda_i \neq \lambda_j \ for \ i \neq j \quad j = 1, 2, \ldots \ldots n.$$

Then:

$$e^{Jt} = \begin{pmatrix} e^{\lambda_1 t} & \ldots & 0 \\ & e^{\lambda_2 t} & : \\ 0 & & e^{\lambda_s t} \end{pmatrix}$$

and $\left\| e^{Jt} \right\| = Max \left| e^{\lambda_1 t} \right| \leq e^{t \max |Re \lambda i|}$

When showing that $\left| e^{At} \right| \leq c e^{\infty t}$, $c \geq 1$. Note the following $\left\| e^{J_0 t} \right\| \leq e^{Bt}$

where . $\beta = \max\limits_{1 \leq l \leq n} |Re \lambda i|$ . Thus $\left\| e^{Jt} \right\| \leq e^{\beta t} e^t, t > 0$

and $\left\| e^{At} \right\| \leq c e^{(\beta+1)t}$   Even more importantly is the following result:

$$\left\| e^{Jt} \right\| \leq \max\limits_{1 \leq i \leq j+k} \left( \left\| e^{J_0 t} \right\|, \left| e^{\lambda_{i+1}} \right| \left\| e^{N_i t} \right\| \right)$$

$$= \max\limits_{1 \leq i \leq n} \left( e^{\beta t}, \left\| e^{\beta t} e^{N_i t} \right\| \right)$$

But $\qquad \left\| e^{\beta t}\, e^{Nnit} \right\| = \sum_{r=0}^{n_1-1} \dfrac{t^r}{r!} e^{\beta t}$ . If $\beta < 0$ But $\beta = -\alpha$, for some $\alpha > 0$

Then:

$$\left\| e^{Jt} \right\| \le \max_i \left( e^{\alpha t}, \sum_{r=0}^{n_1-1} \dfrac{t^r}{r!} e^{-\alpha t} \right)$$

$$\overset{r}{x} = Ax, \quad x(t_0) = x_0$$

Put $x = Py$ then $y = P^{-1}x$.

Therefore, $y = P^{-1}x = P^{-1}Ax = P^{-1}APy = Jy, y(t_0) = P^{-1}x_0$, where $J = P^{-1}AP$.

Here, the solution of initial value problem (IVP) is given by $y(t) = e^{J(t-t_0)}P^{-1}x_0$,

thus, $e^{At} = Pe^{Jt}P^{-1}$.

**Remark 5.3**

If A has distinct eigenvalues $\lambda_1, \lambda_2, \dots \lambda_{ri}$ , then:

$$J = P^{-1}AP = \begin{bmatrix} e^{\lambda_1 t} & 0 & 0 & \cdots & 0 \\ 0 & e^{\lambda_2 t} & 0 & 0 & 0 \\ & & & & e^{\lambda_n t} \end{bmatrix}$$

$$\mathrm{diag}\left[ e^{\lambda_1 t}, e^{\lambda_2 t}, \cdots, e^{\lambda_n t} \right] \tag{5.14}$$

$$e^{At} = P \begin{bmatrix} e^{\lambda_1 t} & & 0 \\ & \ddots & \\ 0 & & e^{\lambda_n t} \end{bmatrix} P^{-1}$$

1. If A has repeated eigenvalues, we can generate $n$ linear independent eigenvectors.

Now:

$$\frac{t^r}{r!}e^{-\alpha t} = \frac{t}{r!e^{\alpha t}} = \frac{t^r}{r!\left(1+\alpha t + \ldots + \dfrac{(\alpha t)^{r+1} r}{(r+1)!} + \ldots\right)}$$

$$< \frac{t^r}{r!(\alpha t)^{r+1}} = \frac{(r+1)!}{(r+1)} \to 0 \text{ as } t \to \infty$$

Since, $\left\|e^{Jt}\right\| \le ce^{-rt}, \alpha > 0$ or in conclusion:

$$\left\|e^{At}\right\| \le ce^{-\alpha t} \text{ for some } \alpha > 0 \tag{5.15}$$

or equivalently:

$$\left\|e^{At}\right\| \to 0, \text{ as } t \to \infty \tag{5.16}$$

The differential equation with these properties is said to be asymptotically stable if it is stable. Detailed treatment is given in the ensuring theorem.

## Theorem 5.3

If all real parts of the eigenvalues of A are negative, then there exists a constant $c \ge 0$ such that:

$$\left\|e^{At}\right\| \le ce^{-rt}, t \ge 0$$

Hence every solution $x(t)$ of equation (5.3) satisfying (5.6) is:

$$\left\|x(t)\right\| \le ce^{-rt}\left\|x_0\right\|, t \ge 0$$

We need to only note that the solution of:

$$\dot{x} = Ax, x(0) = x_0 \tag{5.17}$$

is given by $x(t) = e^{At} x_0$

Hence:

$$\|x(t)\| = \|e^{At}\| \|x_0\|$$
$$\leq c e^{-\alpha t} \|x_0\| \tag{5.18}$$

as required. The reader is further referred to [4] for an alternative proof of the following equivalent result:

**Theorem 5.4**

If all the eigenvalues of A have negative real parts, then every solution $x(t)$ satisfies:

$$\|x(t)\| \leq m e^{-rt} \|x_0\|, \quad \forall t \geq 0 \; for \; some \; m > 0, \; x \geq 0.$$

The theorem can be modified as follows:

**Corollary 5.1**

If all eigenvalues of **A** have negative real parts, every solution $x(t)$ of (5.3) satisfies:

$$\|x(t)\| \leq m e^{-rt} \|x_0\|$$

**Corollary 5.2**

If all real parts of the eigenvalues of A, are at most $\alpha$, then there exists a real number $c \geq 1$ such that:

$$\|e^{At}\| \leq c e^{(\alpha+1)t} \tag{5.19}$$

or every $t$ in E $\forall t \geq 0$ *for some* $m > 0$, $\alpha \geq 0$.

An equivalent result for the scalar situastion is the following Theorem:

## Theorem 5.5

Every solution of the linear autonomous differential equation is:

$$\dot{x} = Ax + h(t) \tag{5.20}$$

where $A$ is $n \times n$ square constant matrix, and $h(t)$ is the $n-$ vector, suppose:

$$\|x(t)\| \leq c \exp(\alpha + ch)(t - t_0)\|x_0\|, x(t_0) = x_0$$

## Proof

By the VCF, the solution of (5.20) is given by:

$$x(t,t_0) = e^{At}\left[ x_0 + \int_{t_0}^{t} e^{-As}h(s)ds \right], t \geq t_0 \tag{5.21}$$

$$\|x(t,t_0,x_0)\| \leq \|e^{At}\|\|x_0\| + \int_{t_0}^{t}\|e^{-A(t-s)}h(s)\|ds$$

$$\leq ce^{-\alpha(t-t_0)}\|x_0\| + \int_{t_0}^{t} cke^{-\alpha(t-t_0)}ds \tag{5.22}$$

Since $\|e^{At}\| \leq ce^{\alpha t}$.

Let:

$$V(t) = x(t,t_0,x_0)e^{-\alpha(t-t_0)} \tag{5.23}$$

Then:

$$\|V(t)\| \leq c\|x_0\| + \int_{t_0}^{t} ck\|V(s)\|ds$$

By Gronwall's Lemma (lemma 5.1):

$$\|V(t)\| \leq c\|V_0\|\exp\left(\int_{t_0}^{t} ck\,ds\right)$$

$$= c\|V_0\|\exp(t - t_0)ck$$

*i.e.*

$$\|x(t,t_0,x_0)e^{-\alpha(t-t_0)}\| \leq c\|V_0\|\exp[ck(t-t_0)]$$

$$\Rightarrow \|x(t,t_0,x_0)\| \leq c\exp(\alpha + ck)(t - t_0)\|x_0\|$$

(5.24)

**Upper bound and Lower bound of Solutions**

Let

$$\dot{x}(t) = A(t)x(t) + f(t),$$

(5.25)

where $A$ is an $n \times n$ matrix-valued function, $f$ is an $n$ vector-valued function whose components are Lebesque summable on $J = [0, \infty)$.

Let $x(t) = x(t, t_0, x_0)$ be the solution of $\dot{x}(t) = A(t)x(t), x(t_0) = x_0$ on the interval $[0, \infty)$.

Define:

$$V(x) = xx^T = \|x\|^2$$

(5.26)

Then:

$$\frac{dV}{dt} = 2\varphi^T(t)A(t)\varphi(t)$$

(5.27)

Thus:

$$-2\|A(t)\|\|\varphi\|^2 \leq \frac{dv(\varphi(t))}{dt} \leq 2\|A(t)\|\|\varphi\|^2$$

or

$$-2\|A(t)\|\|V(\varphi(t))\| \le \frac{dv(\varphi(t))}{dt}$$

$$\le 2\|A(t)\|\|V(\varphi(t))\|$$

Integration yields:

$$\exp\left[-2\left|\int_{t_0}^{t}\|A(s)\,ds\right|\right]V(t_0) \le V(\varphi(t)) \le \exp 2\int_{t_0}^{t}\|A(s)\|\,ds$$

*OR*

$$\exp\left[-2\left|\int_{t_0}^{t}\|A(s)\|\,ds\right|\right]\|x_0\| \le \|x(t,t_0,x_0)\|$$

$$\le \exp\left[\int_{t_0}^{t}\|A(s)\|\,ds\right]\|x_0\|$$

**Theorem 5.6**

Consider a linear system, $\dot{x}(t) = A(t)x(t), x(t_0) = x_0$ where $A(t)$ is an $n \times n$ continuous matrix function of $t$ on $[t_0, +\infty)$. Suppose $M(t), m(t)$ are the maximum and minimum eigenvalues of the symmetric matrix $A(t) + A^*(t)$, where $A^*(t)$ is the transpose of $A(t)$.

The following theorem is imperative:

**Theorem 5.7**

If $x(t)$ is a solution of (5.28) on $(t_0, \infty)$,

Then:

$$0 \le \left\| x(t,t_0,x_0) \right\|^2 \le \left\| x(t_0,t_0,x_0) \right\|^2 \exp\left( \int_{t_0}^{t} m(s)\,ds \right)$$

$$\left\| x(t,t_0,x_0) \right\|^2 \ge \left\| x(t_0,t_0,x_0) \right\|^2 \exp\left( \int_{t_0}^{t} m(s)\,ds \right)$$

Proof

$$\frac{d}{dt}\left| x(t) \right|^2 = \frac{d}{dt} x^* x = \dot{x}^* x + x^* \dot{x}$$
$$= x^* \left( A + A^* \right) x$$

Since $A(t) + A^*(t)$ is symmetric and $M(t)$ is maximal, it follows that:

$$\frac{d}{dt}\left\| x(t) \right\|^2 \le \left\| x^* \left( A + A^* \right) x \right\|$$
$$\le M(t) \left\| x(t) \right\|^2$$

and

$$\frac{d}{dt}\left\| x(t) \right\|^2 \le M(t) \left\| x(t) \right\|^2 .$$

Integration yields:

$$0 \le \left\| x(t) \right\|^2 \le \left\| x_0 \right\|^2 \exp\left( \int_{t_0}^{t} M(s)\,ds \right) \text{ and } \left\| x(t) \right\|^2 \ge \left\| x_0 \right\|^2 \exp\left( \int_{t_0}^{t} m(s)\,ds \right)$$

**Example 5.4**

Consider the system:

$$\dot{x}(t) = \begin{pmatrix} -\dfrac{t}{4} & t^2 \\ -t^2 & -2 \end{pmatrix} x(t), \; t > 0$$

prove that if $x(t,t_0,x_0)$ is the solution of the system through $(t_0\, x_0)$, then:

$$\exp\left(4\left(t-t_0\right)\right)\left\|x_0\right\|^2 \leq \left\|x(t,t_0,x_0)\right\|^2$$

$$\leq \left\|x_0\right\|^2 \sqrt{\frac{t_0}{t}}, \, t > 0$$

**Proof**

$$A(t) + A^*(t) = \begin{pmatrix} -\dfrac{t}{2} & 0 \\ 0 & -4 \end{pmatrix}, \, t > 0$$

$$M(t) = \frac{-1}{2t}, \, m(t) = -4$$

$$\exp\left(\int\limits_{t_0}^{t} m(s)\,ds\right) = e^{In\sqrt{\frac{t_0}{t}}} = \sqrt{\frac{t_0}{t}}$$

$$\exp\left(\int\limits_{t_0}^{t} m(s)\,ds\right) = \exp\left(-4\left(t-t_0\right)\right)$$

Hence:

$$\exp\left(-4\left(t-t_0\right)\right)\left\|x_0\right\|^2 \leq \left\|x(t,t_0,x_0)\right\|^2 \leq \left\|x_0\right\|^2 \sqrt{\frac{t_0}{t}}$$

**Error Bound**

***Theorem 5.8***

Let $x(t), y(t)$ be the solution of $\dfrac{dx(t)}{dt} = f\left(t, x(t)\right), \, t \in [a, b]$

$$f\left(t, x(t)\right) = \begin{pmatrix} f, (t, x) \\ f_2(t, x) \end{pmatrix}, \ D = [a, b] \times E^n \ ,$$

where $f(t, x) \in C^0(D)$ and let $\|f_1(z, t) - f_2(z, t)\| \le \epsilon, z \in D, t \in [t_0, \infty)$

Then:

$$\|x(t) - y(t)\| \le \|x(t_0) - y(t_0)\| e^{L(t - t_0)} + \frac{\epsilon}{L}\left[e^{L(t - t_0)} - 1\right]$$

**Proof**

$$\text{Let } \eta(t) = \|x(t) - y(t)\|^2 = \sum_{k=1}^{n} \|x_k(t) - y_k(t)\|^2$$

Differentiating with respect to and using the Schwartz and triangular inequality,

$$\|\eta(t)\| \le 2\|f(x(t), t) - f(y(t), t)\|\|x(t) - y(t)\|$$
$$+ 2|f_1(x(t), t) - f(y(t), t)|\|x(t) - y(t)\|$$
$$+ 2|f_1(z(t), t) - f(y(t), t)|\|x(t) - y(t)\|$$

Thus:

$$|\eta(t)| \le 2\epsilon \eta(t) + 2\epsilon \sqrt{n(t)}$$

By using $\eta(t) = v^2$, integration yields the proof.

**Problem 5.1**

1.  Let $x(t)$ and $y(t)$ be approximate solutions of the system (5.1) with deviations $\epsilon_1$ and $\epsilon_2$ defined for $a \le t \le b$, *show that*:

$$|x(t) - y(t)| \le |x(a) - y(a)| e^{L(t-a)} + (\epsilon_1 + \epsilon_2)\{e^{L(t-a)} - 1\}L$$

2. Let $F(x,y)$ and $G(x,y)$ be everywhere continuous. Let $F$ satisfy a Lipchitz condition with respect to $t$ and $L$ the Lipchitz constant. Let $\|F(x,y)-G(x,y)\| \le K$. Show that if $f(x)$) and $g(x)$ are approximate solutions of the differential equation:

$$y' = F(x,y)$$
$$z' = G(x,y)$$

with deviations $\in$ *and* $\eta$ *then*:

$$|f(x)-g(x)| \le |f(a)-g(a)|e^{L|x-a|} + (K+\eta+\in)\left\{e^{L|x-a|} -1\right\}L$$

## Continuous Dependence on Initial Data

In the introduction of this chapter, we mentioned the importance of norms of matrices to the concept of continuous dependence on initial data.

This section, highlights the concept of continuous dependence on initial data. The concept is a localized form of a more global concept of continuous dependent on initial parameters from which the stability theory has evolved. The whole lot of chapter thirteen has been devoted to stability.

## Definition 5.1

Let $\dot{x}(t) = A(t)x(t)+h(t), x(t_0)=x_0$ ,where $A(t)$ is an $n \times n$ matrix function of $t$, for $t \in E^1 =(-\infty,+\infty)$.

Suppose the equation has a unique solution, $x(t,t_0,x_0), x(t_0)=x_0$ jointly continuously dependent on initial data if given $\in^* \ge 0$, *and* $\in>0$. Then there exists $\delta(t^*)>0$ such that the solution $x(t,t_0,x_0)$ exists for $\|x_0\| \le \delta(t^*)$ for, $t \in [t_0, t_0 +t^*]$ and $\|x(t,t_0,x_0)\| \to 0$ as $\|x_0\| \to 0$ uniformly. The concept of continuous dependence on initial is a local concept restricted to finite intervals of I, for instance:

$$\dot{x}=kx^2 \quad x(0)=x_0, \quad k \in E^* =(0,\infty) \text{ has a solution } x(t,0,x_0)= -\frac{x_0}{(x_0t -1)}$$

Notice, $x(t,0,x_0)$ is continuous in the domain of definition, and $x(t,0,0)=0$. It follows that given $t^*>0$ and any $\epsilon>0$, there is $S=\delta(\epsilon,\eta^*)$, such that $\|x(t,t_0,x_0)\|<\epsilon$ if $\|x_0\|<S$. Picking $\epsilon=-\dfrac{S}{1-\delta t^*}$ as $t^*$ increases $\delta$ decays, that is, $\delta\to0$ as $t^*\to\infty$.

In conclusion, $x(t,0,x_0)$ is continuous but not uniformly with respect to $x_0$ in the infinite interval $[0,\infty)$. Moreover, if $x(t)=C>0$, no solution even exists.

## Problem 5.2

Estimate from above and from below the solution of $y' = \sqrt{2+y^2-\sin y}$, $y(0)=-3$ and show that the following inequality is valid:

$$\frac{1}{\sqrt{3}}\cosh^{-1}(\sqrt{3}-x) \le y(x) \le \cosh^{-1}(\cosh 3 - x)$$

## Theorem 5.9

Consider the initial value problem:

$$\dot{x}(t) = Ax(t), x(0) = I \tag{5.28}$$

Existing on the interval $[0,+\infty)$ such that $\|A(t)\| \le \eta(t)$

Then:

$$\|x(t)-T\| \le \exp\left(\int_0^x 2sds\right)-1$$

## Proof

$$x(t) = I + \int_0^t A(s)x(s)ds \tag{5.29}$$

$$\|x(t) - I\| \geq \int_0^t \|A(s)\|\|x(s)\| ds \qquad (5.30)$$

$$\phi(t) = \|x(t) - I\| \qquad (5.30)$$

Then

$$\phi(t) + 1 \geq \|x(t)\| \text{ and follows that}$$

$$\phi(t) \leq \int_0^x \eta(s)(\phi(s) + 1) ds \qquad (5.32)$$

Let:

$$w(t) = 1 + \phi(t)$$
$$w(t) = \phi(t)$$

But:

$$\phi(t) \ w(t) \leq \eta(t)|1 + \phi(t)| = \eta(t)w(t) \qquad (5.33)$$

Integration given:

$$w(x) \leq w(0)\exp\left(\int_0^x \eta(s)ds\right)$$

$$= 1.\exp\left(\int_0^x \eta(s)ds\right)$$

$$\|x(t) - I\| + 1 \leq \exp\left(\int_0^x \eta(s)ds\right)$$

$$\|x(t) - I\| \leq \exp\left(\int_0^x \eta(s)ds\right) - 1$$

## Linear Periodic Systems

In this section, linear periodic systems are studied; Floquet rule is applied to find solutions to linear periodic systems. Theorems on how to construct monodromy matrices for linear periodic systems [8] are considered with examples given.

## NOTATION

Let:

$p_T$ is the space of T – periodic function $L_p$ linear periodic systems.

$R_p$ is space of solutions of reciprocal systems. The reciprocal systems will be introduced later.

Consider the Linear System:

$$\dot{x} = A(t)x(t), \tag{5.34}$$

where $A(t+T) = A(t)$ and $A(t+T^*) \neq A(t)$ for any other $T^* < T$ . Then the system is called a T- periodic system. Geometrically, it has a property with its graph shifts T units to left (or right) leaving it unaltered. We will see later on that a system that admits periodic solutions often has their solutions as the product of $e^{Bt}$ and a periodic function $P(t)$ for some constant matrix $B$ .

Examples of periodic systems are found in nature with periodic external force (often referred to as simple harmonic motion).Similarly, in inductance, capacitors and resistors (LCR) circuits practically demonstrate periodicity.

There are many classes of linear periodic systems in $L_p$ . Hill's and Mathieu equations are typical examples. Some systems are composed of sums of periodic systems; they are referred to as almost periodic systems. We will not study this class of the $L_p$ systems in this chapter.

In Hill's equations, the companion matrix, $A(t)$ is of the form:

$$A(t) = \begin{pmatrix} 0 & 1 \\ -\lambda - \phi(t) & 0 \end{pmatrix}$$

where $\phi(t)$ is T-periodic, real and continuous function and $\lambda$ is a constant.

The Mathieu equation, on the other hand, is a particular case of the Hill's equation, in which case $\phi(t) = \mu \cos 2t$, $\mu =$ constant.

The core of our investigation will be on stability properties of $L_p$, which will be thoroughly discussed.

The fundamental matrix solution in $L_p$ if exists is of the form $X(t) = R(t)e^{Bt}$, as stated above, where $X(t)$ is T-periodic and $B$ is a constant matrix. This is the renowned Floquent rule. Nevertheless, before the proof will be outlined, it is pertinent to claim the existence of a non-singular n-matrix $C = e^B$. Matrix C is nothing other the Jordan canonical block form. For more information (See Jack [8], pp. 118).

We consider the celebrated Floquent rule:

**Theorem 5.10** (The Floquent rule)

Every fundamental solution $x(t) \in L_p$ is of the form:

$$x(t) = R(t)e^{Bt} \tag{5.35}$$

where $R(t+T) = R(t)$, for every $t$, where $B$ is a constant matrix.

**Proof**

Let $X(t)$ be the fundamental matrix of Lp . We will be done if only we ended showing that $R(t)$ is T-periodic.

$$X(t+T) = X(t)C$$
$$= R(t)e^{Bt}$$

$$\text{But, } R(t) = X(t)e^{-Bt}$$

Hence:

$$R(t+T) = X(t+T)e^{-B(t+T)}$$
$$= X(t+T)e^{-Bt}e^{-Bt}$$
$$= X(t)e^{-Bt}e^{-Bt} \tag{5.36}$$
$$= X(t)e^{-Bt} = R(t)$$

## Definition 5.2

A monodromy matrix of $L_P$ is an invertible matrix C associated with a fundamental solution $x(t)$ of $L_P$ for $X(t+T) = X(t)C$ . The eigenvalues $\lambda_j$ of the monodromy matrix $C$ are called the characteristic multipliers of $L^P$ whereas, $P_j = e^{\lambda_j T}, j = 1, 2, .. n$ is the characteristic exponent.

Another class of $L^P$ is the reciprocal system $R^P$, for these systems, both J and $P^{-1}$ are characteristic multipliers of complex functions in $R_p$ .

## Theorem 5.11

If $P_j$ are the characteristic multipliers of $L_p$ for the index $j = 1, 2, \cdots$ ,then:

$$\prod_{j=1}^{n} P_j = \exp\left( \int_0^T TrA(s)ds \right) \tag{5.37}$$

$$\sum_{i=1}^{n} \lambda_j = \frac{1}{T}\left( \int_0^T TrA(s)ds \right)(\mod \frac{2\pi i}{T}) \tag{5.38}$$

Tr $A(.)$ is the trace of matrix $A(.)$ .

## Proof

Let C be the monodromy matrix for the principal matrix solution, $X(t), X(0) = 1$ .

By Liouvilles' Theorem, $\det C = \det X(T) = \exp\left( \int_0^t TrA(s)ds \right)$

The proof follows immediately from the definition.

## Theorem 5.12

Let $X(t)$ be the fundamental matrix solution of $L_p$:

$$X(t+T) = X(t)C \qquad (5.39)$$

Let $Y(t)$ be another, then the monodromy matrix of $X(t)$ and $Y(t)$ are similar matrices.

## Proof

$Y(t)$ is the fundamental matrix. By hypothesis, there exists a non-singular matrix D such that:

$$Y(t+T) = Y(t)D \qquad (5.40)$$

See the statement underneath the theorem.

$X(t)X(t) = Y(t)C$ for some non-singular matrix $C$.

$$X(t+T) = Y(t+T)C = Y(t)DC = X(t)C^{-1}DC$$

$$X(t+T) = X(t)C^{-1}DC = X(t)E .$$

Where,

$$E = C^{-1}DC \qquad (5.41)$$

Hence, monodromy matrix E is similar to the monodromy matrix D.

## Corollary 5.3

Any two monodromy matrices of a linear periodic system have same eigenvalues and hence some characteristic exponents.

**Proof**

The proof is trivial since, similar matrices have same eigenvalues (See Goel [20], pp.187).

The proof is immediate from definition.

**Example 5.5**

Show that the system

$$\dot{x}(t) = \begin{pmatrix} -1 + \dfrac{3}{4} Cos^2 t & 1 - \dfrac{3}{4} Cost\, S\, int \\ -1 - \dfrac{3}{4} S\, int\, Cost & 1 + \dfrac{3}{4} Sin^2 t \end{pmatrix} x(t)$$

is periodic with period $T = 2\pi$. Will the system admit a periodic solution?

**Solution**

$$A(t) = \begin{pmatrix} -1 + \dfrac{3}{4} Cos^2 t & 1 - \dfrac{3}{4} Cost\, S\, int \\ -1 - \dfrac{3}{4} S\, int\, Cost & 1 + \dfrac{3}{4} Sin^2 t \end{pmatrix} x(t)$$

$A(t + 2\pi) = A(t)$, since $\sin(t + 2\pi) = \sin t, \cos(t + 2\pi) = \cos t$.

$$\sin^2 t = \frac{1}{2}(1 + \cos^2 t), \sin^2 t = \frac{1}{2}(1 - \cos^2 t).$$

Thus, $A(t)$ is $2\pi$ periodic and by Floquent rule, it admits a periodic solution.

**Example 5.6**

Let $a_0(t), a_1(t)$ be continuous and $w$ – periodic functions and $\phi_1, \phi_2$ be the solutions of:

$$\ddot{y} + a(t)\dot{y} + a_o(t)y = o$$

such that the matrix product X satisfies:

$$X(0) = \begin{vmatrix} \phi_{11}(0) & \phi_{21}(0) \\ \phi_{21}(0) & \phi_{22}(0) \end{vmatrix} = \begin{bmatrix} 1 & 0 \\ 0 & 1 \end{bmatrix} = I$$

Show that the characteristic multipliers satisfy:

$$\lambda^2 + \alpha\lambda + \beta = 0,$$

where

$$\alpha = \phi_1(w) + \phi_2(w)$$

$$\beta = \exp\left(\int_0^w -a_1(s)ds\right)$$

If one characteristic multiplier has had:

$$\frac{1}{100}\exp\left(\int_o^w -a_1(s)ds\right)$$

What is the value of the other multiplier?

**Solution**

Let:

$$\dot{y} = y_1$$

$$\dot{y}_1 = y_2 = -a_0(t)y - a_1(t)y_1$$

$\mathrm{Tr}A(t) = -a_1(t)$. The characteristic multiplier is the root of $\det[X(\omega) - p] = 0$

$$\Rightarrow \lambda^2 - Tr\ A(t)\lambda + \det X(\omega)$$
$$= \lambda^2 - \left[\phi_{11}(0) + \phi_{22}(0)\right]\lambda + \det X(\omega)$$

$$\det X(\omega) = \phi_{11}\phi_{22} - \phi_{12}\phi_{21} = \exp\left(\int_0^\omega -a_1(s)ds\right)$$

Therefore, $\lambda^2 + \alpha\lambda + \beta = 0$ implies that: $\alpha = -\left[\phi_{11}(0) + \phi_{22}(0)\right], \beta = \exp\left(\int_{t_0}^{t} a_1(s)ds\right)$,

but $\det X(\omega) = 1, \lambda_1\lambda_2 = 1$.

One of the characteristic multipliers has the value $\lambda_1 = \dfrac{1}{100}\exp\left(\int_0^{\omega} a_1(s)ds\right)$ and

hence the second one is $\lambda_2 = 100\exp\left(\int_0^{\omega} a_1(s)ds\right)$.

## CONCLUSION

In this chapter, the following are introduced: Theory of autonomous linear homogeneous systems introduced together with the canonical transform process and Jordan Canonical form. The Sylvester formula for constructing fundamental matrix solution and estimating bounds for solution, along with the method for obtaining an upper and lower bound of solution to ODES has been obtained. The concept of solution depends on initial data, linear periodic system and application to some examples made.

## REFERENCES

[1]    W.E. Boyce, and R.C. Diprima, *Elementary Differential Equations and Boundary Value Problems,* 7th ed John Wiley and Son, 1977.

[2]    B.H. Chirgwin, and C. Plumpton, *Advanced Theoretical Mechanics: A Course of Mathematics..* Pergamon press Ltd, 1961.

[3]    B.H. Chirgwin, and C. Plumpton, *A Course of Mathematics for Engineers and Scientists..* Pergamon press Ltd., 1961.

[4]    E.A. Coddington, and N. Levinson, *Theory of Ordinary Differential Equations.* McGraw Hill, 1955.

[5]    E. Kreyszig, *Advanced Engineering Mathematics.* John Wiley Publication: USA, 2000.

[6]    G. Birkhoff, and G-C. Rota, *Ordinary Differential Equations.* Wiley, 1989.

[7]    A.D. Mysvkis, *Advanced Mathematics for Engineers: Special Courses.,* 2nd ed MIR Publication Moscow, 1979.

[8]    J.K. Hale, *Ordinary Differential Equations.* Wiley-Interscience: New York, London, Sydney, Toronto, 1969.

# CHAPTER 6

# Stability Theory

**Abstract:** Stability including its characterizations is considered using quantitative and qualitative theories. Stability criteria are discussed using the Routh-Hurwitz criterion and fundamental matrices. Stability is investigated for nonlinear systems through linearization and Lyaponov's methods. Applications are made to single and multi-species population models.

**Keywords:** Fundamental matrices, Linearization, Lyaponov's stability, Multi-species populations, Nonlinear systems, Routh-Hurwitz criteria, Stability criteria, Single species populations.

## INTRODUCTION

Stability over the century has constituted the backbone of study for modern dynamical systems. Scientists and engineers often take this concept into consideration whenever a mechanism is to be designed [1,2,3,5].

Intuitively, the concept could be said to have evolved from the study of the behavior of a system when perturbed (disturbed) from its equilibrium (resting) positions when the motion would not radically deviate from resting positions; for example, the vibration of a simple pendulum, when displaced from the equilibrium position in such a way that the amplitude of the oscillation is small.

The motion of a ball on a smooth parabolic surface is a motion exhibiting stability phenomenon [10].

### Stability Theorems

Consider the initial value problem (IVP):

$$\dot{x}(t) = f(t, x(t)), x(t_0) = x_0 \qquad (6.1)$$

Where, $f \in C(I \times \Omega, E^n)$, $\Omega$ is an open and connected subset of $E^n$. Assuming $f$ has the property that the solution $x(t, t_0, x_0)$ exists and is unique, for example, if

**Benjamin Oyediran Oyelami**
**All rights reserved-© 2024 Bentham Science Publishers**

$x(t,t_0,x_0)$ is a solution and $\iota$ satisfies the hypotheses of the Picard-Linderlof existence and unique theorem (see Chapter 10) .

The core or central question of stability is as follows: is there a special solution $\phi(t,t_0,x_0)$ existing on the interval $[t_0,+\infty)$ such that a small perturbation would result in small deviations from system behavior? If such a solution exists, we say such a system is stable, otherwise it is unstable [7-10].

Stability treatment covers a vast majority of phenomena. For the purpose of a comprehensive study, our scope of discussion will encompass three considerations, namely stability through $(\varepsilon - \delta)$ argument. The interpretation of this will be given when the fundamental definition of stability is stated. Secondly, we will consider stability *via* fundamental matrices and finally, by the use of a scalar function called the Lyapunov stability function [4-7]. This method is fundamentally based on the energy concept.

## Stability and its Characterization by $(\varepsilon - \delta)$ Argument

## Definition 6.1

Let $\phi(t) = \phi(t,t_0,y_0)$ be a solution of (6.1) such that $\phi(t_0) = y_0$ . The solution $\phi(t)$ is said to be stable (in Lyapunov sense, L. S.), if given $\in > 0, t_0 \in E^1$ , there exists $\delta(t,\in,t_0) > 0$ , such that $\|x_0 - y_0\| < \delta$ implies that: $\|x(t,t_0,x_0) - \phi(t)\| < \in, t \ge t_0 + T(\eta)$ , for every $t \ge t_0$ ; otherwise unstable $\phi(t)$ is asymptotically stable if it is stable and in addition:

$$\|x(t,t_0,x_0) - \phi(t)\| \to 0 \text{ as } t \to \infty \tag{6.2}$$

$\phi(t)$ is uniformly stable, if the choice of $\delta(t,t_0,\in)$ is independent of $t_0$ or equivalently, for every $\eta > 0$ , there exists $T(\eta)$ such that $\|x_0 - y_0\| < \delta$ implies that $\|x(t,t_0,x_0) - \phi(t)\| < \in, t \ge t_0 + T(\eta)$ .

Fig. (**6.1**) gives a diagrammatic explanation of the Stability concept which shows that a system is stable if the trajectory of the solution remains in the circle $S_1$ with a small radius $\epsilon > 0$ for a given circle $S_2$ with radius $\delta > 0$ containing the initial data of the system. The trajectory will never penetrate the boundary of $S_1$ ($\|x(t)\| < \epsilon, t \geq t_0$) for the finite value of norm of the initial data ($\|x_0\| < \delta$).

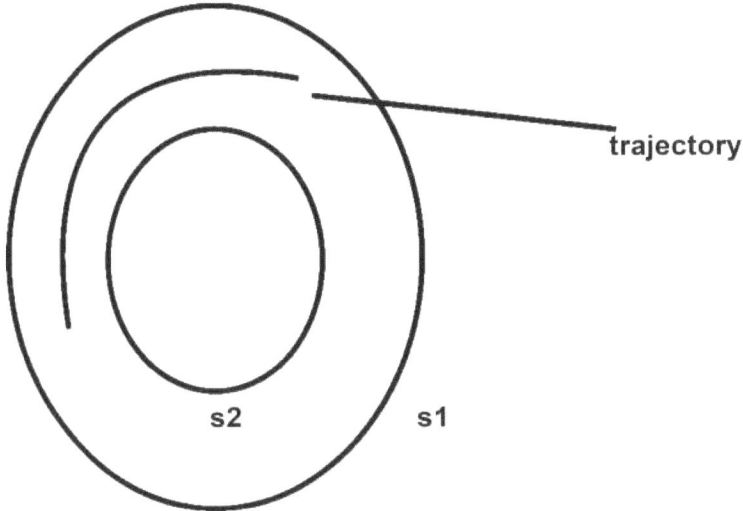

trajectory

s2          s1

**Fig. (6.1) Stability definition.**

## Geometric Interpretation

A solution $x(t, t_0, x_0)$ is called a trajectory of motion in Definition 6.1. This is interpreted as follows: The origin 0 is stable if given $\epsilon > 0$ and:

$$S_1 = \{x \in E^n : \|x\| < \epsilon\} \tag{6.3}$$

there exists a ball $S_2$ where:

$$S_2 = \{x \in E^n : \|x\| < \delta\}, \ \delta < \epsilon \ \text{in} \ E^n, \tag{6.4}$$

such that the trajectory will never penetrate S2.

For asymptotic stability, if conditions in Definition 6.1 hold, then $x(t, t_0, x_0) \to 0$ as $t \to \infty$. Stability can be geometrically interpreted in terms of trajectories remaining in the cylinder of infinite radius $\epsilon$.

In Engineering, asymptotic stability is desirable since the solution eventually decays to zero.

The implication of Theorem is that equation (6.1) is asymptotic stable,

since $\|x(t)\| \le \|e^{At}x_0\| \le \|e^{At}\|\|x_0\| \le Ce^{\alpha t}\|x_0\| \to 0$ as $t \to \infty$, as long as all real parts of the eigenvalues of A are negative.

## Routh-Hurwitz Criteria

The Routh-Hurtwiz criterion is a method for determining stability for continuous systems with the characteristic equation of the form: $a_n\lambda^n + a_{n-1}\lambda^{n-1} + \ldots + a_1\lambda + a_0 = 0$ (6.5)

The criterion is applied using the following Table **6.1**.

**Table 6.1. Routh-Hurwitz Table.**

| $s^n$ | $a_n$ | $a_{n-2}$ | $a_{n-4}$ | $a_{n-6}$ | $\ldots$ |
|---|---|---|---|---|---|
| $s^{n-1}$ | $a_{n-1}$ | $a_{n-3}$ | $a_{n-5}$ | $a_{n-7}$ | $\ldots$ |
| $\cdot$ | $b_1$ | $b_2$ | $b_3$ | $b_4$ | $\ldots$ |
| $\cdot$ | $c_1$ | $c_2$ | $c_3$ | $c_4$ | $\ldots$ |
| $\ldots\ldots$ | $\ldots\ldots$ | $\ldots\ldots$ | $\ldots$ | $\ldots$ | $\ldots$ |

where $a_n, a_{n-1}, \ldots, a_n$ are the coefficients of the characteristic equation such that:

$$b_1 = \frac{a_{n-1}a_{n-2} - a_n a_{n-3}}{a_{n-1}} \quad b_1 = \frac{a_{n-1}a_{n-4} - a_n a_{n-5}}{a_{n-1}}$$

$$c_1 = \frac{b_1 a_{n-3} - a_{n-1}b_2}{b_1} \quad c_2 = \frac{b_1 a_{n-5} - a_{n-1}b_3}{b_1}$$

The table is continued horizontally and vertically until zeroes are obtained.

The Routh Criterion: All the roots of the characteristic equation have negative real parts if and only if the elements of the first column of the Routh table have the same sign. Otherwise, the number of roots with positive real parts is equal to the number of changes of sign.

## Remark 6.1

This criterion is equivalent to saying that a differential equation whose characteristic equation is the equation (6.5) is asymptotically stable if all real parts of the eigenvalues are negative and unstable if at least one eigenvalue of it has a positive real part.

## Hurwitz Criterion

All the real roots of the characteristic equation have real parts if and only if
$D_k > 0, k = 1, 2, 3, \cdots$

Where:

$$D_1 = a_{n-1}$$

$$D_2 = \det \begin{bmatrix} a_{n-1} & a_{n-3} \\ a_n & a_{n-2} \end{bmatrix} = a_{n-1}a_{n-2} - a_n a_{n-3}$$

$$D_3 = \det \begin{bmatrix} a_{n-1} & a_{n-3} & a_{n-5} \\ a_n & a_{n-2} & a_{n-4} \\ 0 & a_{n-1} & a_{n-3} \end{bmatrix} = a_{n-1}a_{n-2}a_{n-3} - a_n a_{n-1}a_{n-5} - a_n a_{n-3}^2 - a_{n-4}a_{n-1}^2$$

$$\vdots$$

$$D_k = \det \begin{bmatrix} a_{n-1} & a_{n-2} & \cdots & \begin{bmatrix} a_0 & \text{if } n \text{ odd} \\ a_1 & \text{if } n \text{ even} \end{bmatrix} & 0 & 0 & \cdots & 0 \\ a_n & a_{n-2} & \cdots & \begin{bmatrix} a_1 & \text{if } n \text{ odd} \\ a_0 & \text{if } n \text{ even} \end{bmatrix} & 0 & 0 & \cdots & 0 \\ 0 & a_{n-1} & a_{n-3} & \cdots & & \cdots & \cdots & \cdots & \cdots \\ 0 & a_n & a_{n-2} & \cdots & & \cdots & \cdots & \cdots & 0 \\ \cdots & \cdots & \cdots & \cdots & & \cdots & \cdots & \cdots & \cdots \\ 0 & \cdots & \cdots & \cdots & & \cdots & \cdots & \cdots & a_0 \end{bmatrix}$$

## Example 6.1

Investigate the stability of the following differential equation:

$$\frac{d^3 y}{dx} + 6\frac{d^2 y}{dx} + 12\frac{dy}{dx} + 8 = 0 \tag{6.6}$$

## Solution

The characteristic equation for the differential equation is:

$$\lambda^3 + 6\lambda^2 + 12\lambda + 8 = 0$$

Applying Routh Table, we have the following Table **6.2**.

**Table 6.2. Routh criterion.**

| | | | |
|---|---|---|---|
| $\lambda^3$ | 1 | 12 | 0 |
| $\lambda^2$ | 6 | 8 | 0 |
| $\lambda$ | $\dfrac{64}{6}$ | 0 | - |

(Table 6.2) cont.....

| $\lambda^0$ | 8 | - | - |
|---|---|---|---|

In Table **6.2**, there are no changes in sign in the first column; therefore, by Routh criterion, all the roots of the equation have negative real parts.

## Example 6.2

Investigate the stability of the following differential equation:

$$\frac{d^3 y}{dx} + 3\frac{d^2 y}{dx} + 3\frac{dy}{dx} + 1 + r = 0 \tag{6.7}$$

Applying the Routh Table, we have the following Table **6.3**.

| $\lambda^3$ | 1 | 3 | 0 |
|---|---|---|---|
| $\lambda^2$ | 3 | $1+r$ | 0 |
| $\lambda$ | $\dfrac{8-r}{3}$ | 0 | |
| $\lambda^0$ | $1+r$ | | |

For the system to be stable, all real parts of the eigenvalues must be negative, therefore there should be no change of sign in the first column in Table **6.3**. Hence, the condition for this to happen is that $0 < 8-k, 0 < 1+k$, which implies that $-1 < r < 8$ must be satisfied for the system to be stable.

## *Importance of Hurwitz Criterion*

It gives an algebraic procedure for examining the root of L (P) without actually solving them. The qualitative properties of the solution make it widely applicable in stability theory for scalar equations. However, we consider the other side of the coin.

## Demerit

It does not state the degree of stability or instability of the method is also handicapped for the case where $k \geq 4$, since the evaluation of $D_k$ in this case becomes difficult in practice.

## Stability through Fundamental Matrices

Suppose we have a perturbed linear system:

$$\dot{x}(t) = A(t)x(t) + f(t, x(t)) \tag{6.8}$$

Where $A(t) \in C^0(E^n, E^{n \times n})$. The stability of (6.8) can be studied using the following theorem:

## Theorem 6.1

Let $X(t)$ be the fundamental matrix solution of the homogeneous part of equation (6.8). $X(t)$ is stable for any $t_0 \in E^1$ if and only if there exists $k = k(\beta)$ such that:

(i) $\|X(t)\| \leq k, t_0 \leq t < \infty$, uniformly stable if and only if $\|X(t)X^{-1}(s)\| \leq k, t_0 \leq s \leq t < \infty$.

(ii) Asymptotic stable for any $t_0 \in E^1$ if and only if $\|x(t, t_0, x_0)\| \to 0$ as $t \to \infty$

(iii) Uniformly asymptotic stable for $t_0 \geq \beta$ if only if $\|X(t)X^{-1}(s)\| \leq ke^{-\alpha(t-s)}, \beta \leq s \leq t < \infty$

## Theorem 6.1

$$\text{Let } \dot{x}(t) = A(t)x(t), x(t_0, t_0, x_0) = x_0 \tag{6.9}$$

The system (6.9) is uniformly stable if $\int_{t_0}^{t} \eta(s)ds < M$ for some positive constant such that $\eta(t) \geq \|A(t)\|$.

## Proof

$x(t,t_0,x_0) = x_0 + \int\limits_{t_0}^{t} A(s)x(s)ds, t \geq t_0$ Let $\phi(t,t_0,x_0)$ be the special solution existing on I when (6.8) is perturbed.

Then,
$$\phi(t,t_0,x_0) = \phi_0 + \int\limits_{t_0}^{t} A(s)\phi(s)ds, \phi(t,t_0,x_0) = \phi_0$$

Thus,
$$\left\| x(t,t_0,x_0) - \phi(t,t_0,x_0) \right\| \leq \left\| x_0 - \phi_0 \right\| + \int\limits_{t_0}^{t} \left\| A(s) \right\| \left\| x(s) - \phi(s) \right\| ds$$

$$\text{Let } \psi(t) = x(t) - y(t) \tag{6.10}$$

Then,
$$\left\| \psi(t) \right\| \leq \left\| \psi_0 \right\| + \int\limits_{t_0}^{t} \eta(s) \left\| \psi(s) \right\| ds .$$

Using Gronwall's inequality:

$$\leq \left\| \psi_0 \right\| \exp(\int\limits_{t_0}^{t} \eta(s)ds) \tag{6.11}$$

*i.e.*

$$\left\| x(t,t_0,x_0) - \phi(t,t_0,x_0) \right\| \leq \left\| x_0 - y_0 \right\| \exp(\int\limits_{t_0}^{t} \eta(s)ds) < \delta \exp(\int\limits_{t_0}^{t} \eta(s)ds) \tag{6.12}$$

Given $\varepsilon > 0$, we can find $\delta(\in,t_0,x_0) > 0$ such that:

$$\left\| x_0 - \phi_0 \right\| < \delta \Rightarrow \left\| x_0 - \phi_0 \right\| \exp(\int\limits_{t_0}^{t} \eta(s)ds) < \delta \exp(\int\limits_{t_0}^{t} \eta(s)ds) < \in$$

That is:

$$\delta \exp(\int_{t_0}^{t} \eta(s)ds) < \in$$

$$\Rightarrow \int_{t_0}^{t} \eta(s)ds < \ln(\frac{\in}{\delta}) = M$$

Picking $\delta(\in, t_0, x_0) = \in \exp(-\int_{t_0}^{t} \eta(s)ds)$. The stability is uniform since $\delta$ it is independent of $x_0$. But stability is not asymptotic except when $\lim_{t \to \infty} \int_{t_0}^{t} \eta(s)ds = -\infty$

This is impossible!

Stability and asymptotic stability of any other solution $y(t)$ may be investigated by replacing $x$ by $z + y$ and discussing the zero solution of the equation:

$$\dot{y} = f(t, z + y) - f(t, y) \tag{6.13}$$

The stability of (6.13) is the same for arbitrary solution $y$. Throughout the this study, concentration of our attention will be on the stability of trivial solutions (singular point) rather than the whole system, since both are equivalent in the stability sense.

**Maple Examples (PolynomialTools [Hurwitz] in Help in the Maple 2023)**

The RouthTable command returns the Routh table of the polynomial $p$ as a Matrix. The parameter s is the indeterminate of the polynomial $p$. The table can be used to determine the number of roots of $p$ in either the open right half complex plane (open RHP) or the open left half complex plane (open LHP). If the option StableCondition=true is included, the RouthTable command outputs an expression giving conditions under which the polynomial is stable. In this case, no Matrix is returned.

The Hurwitz ($p$, z) function determines whether the polynomial $p(z)$ has all its zeros strictly in the left half plane. A polynomial is a Hurwitz polynomial if all its roots are in the left half plane.

The parameter $p$ is a polynomial with complex coefficients. The polynomial may have symbolic parameters, which <u>evalc</u> and Hurwitz assume to be real. The paraconjugate $p^*$ of $p$ is defined as the polynomial whose roots are the roots of $p$ reflected across the imaginary axis.

> $with(DynamicSystems)$ :

> $p := (s^2 + 1) (s^2 - 1) (s + 2)$ :

> $RouthTable(p, s)$

$$\begin{bmatrix} 1 & 0 & -1 & s^5 \\ 2 & 0 & -2 & s^4 \\ 8 & 0 & 0 & [s^3] \\ 2 & -2 & 0 & s^2 \\ 8 & 0 & 0 & s \\ -2 & 0 & 0 & 1 \end{bmatrix}$$

There is one sign change in the first column; therefore, there is one root in the open RHP. The $[s^3]$ indicates a degenerate polynomial. Consequently, there might be roots on the imaginary axis. Check the open LHP.

> $RouthTable(p, s, left)$

$$\begin{bmatrix} 1 & 0 & -1 & s^5 \\ -2 & 0 & 2 & s^4 \\ -8 & 0 & 0 & [s^3] \\ -2 & 2 & 0 & s^2 \\ -8 & 0 & 0 & s \\ 2 & 0 & 0 & 1 \end{bmatrix}$$

There are two sign changes in the first column; therefore, there are two roots in the open LHP. Together with the previously determined root in the RHP, this accounts for three roots of this polynomial of degree five, leaving two roots on the imaginary axis.

> $RouthTable(ax^2 + bx + c, x, 'stablecondition' = true)$

*true*

> $RouthTable\left(x^2 + bx + c, x, \text{'stablecondition'} = true\right)$

*false*

> $RouthTable\left(x^2 - 2x + c, x, \text{'stablecondition'} = true\right)$

*false*

> $RouthTable\left(x^2 + bx + a, x, \text{'stablecondition'} = true\right)$

$$0 < a \textbf{ and } 0 < b$$

> *restart*

> $with(PolynomialTools):$

> 
$$p1 := z^2 + z + 1$$

$$p1 := z^2 + z + 1$$

> $Hurwitz(p1, z)$

*true*

> 
$$p2 := 3z^3 + 2z^2 + z + 2c$$

$$p2 := 3z^3 + 2z^2 + 2c + z$$

> $Hurwitz(p2, z, 's2', 'g2')$

*FAIL*

> $s2$

$$\left[\frac{3}{2} z, -\frac{2z}{3c-1}, -\frac{3}{2} z + \frac{1}{2}\frac{z}{c}\right]$$

> $g2$

1

The elements of *s2* are all positive if and only if $0 < c < \frac{1}{3}$, by inspection. Thus, you can use the information returned even when the direct call to Hurwitz fails.

Separate calls to Hurwitz in cases $c = 0$ and $c = \frac{1}{3}$ give nontrivial gcds between *p2* and its paraconjugate. Thus, the stability criteria are satisfied only as above.

>
$$p3 := 4z^4 + z^3 + z^2 + c$$

$$p3 := 4z^4 + z^3 + z^2 + c$$

> *Hurwitz(p3, z, 's3', 'g3')*

*FAIL*

> *s3*

$$\left[ 0, 4z, z, -\frac{z}{c}, -z \right]$$

Notice that the last term has coefficient $-1$. Thus, you can say unequivocally that *p3* is not Hurwitz, for any value of *c*.

>
$$p4 := z^5 + 5z^4 + 4z^3 + 3z^2 + 2z + c$$

$$p4 := z^5 + 5z^4 + 4z^3 + 3z^2 + c + 2z$$

> *Hurwitz(p4, z, 's4', 'g4')*

*FAIL*

> *s4*

$$\left[ \frac{1}{5}z, \frac{25}{17}z, \frac{289}{5}z, \frac{z}{5c+1}, -\frac{1}{17}\frac{(5c+1)^2 z}{c^2 + 48c - 2}, \frac{(-c^2 - 48c + 2)z}{(5c+1)c} \right]$$

By inspecting $s4$, notice that $p4$ is Hurwitz only if $-\frac{1}{5} < c$, and $c^2 + 48\,c < 2$, and $0 < c$. This can be simplified to the conditions $0 < c < -24 + 17\sqrt{2} = 0.04...$

> $$p5 := p2 + 1d$$

$$p5 := 3\,z^3 + 2\,z^2 + c + z + 1d$$

evalc and the Hurwitz function assume that symbolic parameters have real values.

> *Hurwitz*$(p5, z, 's5', 'g5')$

*FAIL*

> $s5$

$$\left[\frac{3}{2}\,z, \; -\frac{4z}{3c-2} - \frac{81d}{(3c-2)^2}, \; -\frac{1}{2}\,\frac{(3c-2)^3\,z}{9c^3 - 12\,c^2 - 8\,d^2 + 4c}\right.$$
$$\left. + \frac{1d\,(3c-2)^2}{9c^3 - 12\,c^2 - 8\,d^2 + 4c}\right]$$

The coefficients of $s5$ are inspected according to rules, but it is a tedious process.

> $$p6 := expand\big((x-1)\,(x^2+2)\,(x-c)\big)$$

$$p6 := -c\,x^3 + x^4 + c\,x^2 - x^3 - 2\,cx + 2\,x^2 + 2\,c - 2\,x$$

> *Hurwitz*$(p6, x, 's6', 'g6')$

*false*

> $g6$

$$x^2 + 2$$

> $$p7 := x + \sqrt{2}$$

$$p7 := x + \sqrt{2}$$

> *Hurwitz*(p7, x)

*true*

>
$$p8 := x^3 + cx^2 + (c^2 - 1)x + 1$$

$$p8 := x^3 + cx^2 + (c^2 - 1)x + 1$$

> *Hurwitz*(p8, x, 's8', 'g8')

*FAIL*

> s8

$$\left[ \frac{x}{c}, \frac{c^2 x}{c^3 - c - 1}, c^2 x - x - \frac{x}{c} \right]$$

Examination of the above for real values of $c$ is a way to determine whether the polynomial is Hurwitz.

>
$$p9 := expand\left( (cz^2 + 1)(z + 1)(z^2 + 2z + 2) \right)$$

$$p9 := cz^5 + 3cz^4 + 4cz^3 + 2cz^2 + z^3 + 3z^2 + 4z + 2$$

> *Hurwitz*(p9, z, 's9', 'g9')

*FAIL*

> s9

[ ]

> g9

$cz^2 + 1$

In the previous example, $c$ might be zero. Thus, Hurwitz cannot determine whether all the zeros are in the left half plane.

## CONCLUSION

The stability theorem from the quantitative point of view is considered in this chapter. Stability and its characterization are studied using (epsilon, delta) argument, Routh-Hurwitz and Hurwitz Criteria and conclusion is made together with examples given through fundamental matrices. Maple software is demonstrated to investigate the systems stability with both Routh-Hurwitz and Hurwitz criteria.

## REFERENCES

[1]  O. Akinyele, and P. Edet Akpan, "On the -Stability", *J. Math. Anal. Appl.,* vol. 164, no. 2, pp. 307-324, 1992.

[2]  J.A. Anderson, G.E. Hilton, Eds., *Parallel Models.* Lawrence Eribaum and Associates: Hillsdale, New York, 1981.

[3]  E. Beltrami, *Mathematics for Dynamic Modelling..* London Academic Press, 1987.

[4]  B.S. Goh, *Stability of some Multispecies Population Models in Modeling and Differential Equations in Biology.* Lecture Notes in Pure and Applied Mathematics Marcel Dekker,Inc. Publication: New York, 1980, pp. 209-216.

[5]  R.M. May, *Stability.* Princeton University Press: Princeton, NJ, 1974.

[6]  V.G. Matsenko, and V.N. Rubainovski, "Application of lyapunov direct method for analyzing the age structure of biological population", *Zhvyscist Math. Fiz,* vol. 23, no. 2, pp. 320-332, 1983.

[7]  B. O. Oyelami, *Studies in Impulsive Systems and Applications.* Lambert Academic Publisher Germany, 2012.

[8]  G. Deo Sadashiv, and V Ragavendra, *Ordinary Differential Equations.* Tata McGraw-Hill: India, 1980.

[9]  G. Deo Sadashiv, V Lakshimikantham, and V Ragavendra, *Textbook of Ordinary Differential Equations..* Tata McGraw-Hill: India, 1997.

[10]  Barnett Stephen, *Cameron..* Clarendon Press: USA, 1985.

# CHAPTER 7

# Stability of Perturbed Systems

**Abstract:** The stability property is investigated for perturbed linear autonomous and non-linear systems using linearization and Lyapunov's methods. Some examples are given on the stability of some nonlinear systems through eigenvalues of the linearized systems and coupled with the estimation of the norm of the error of approximation.

**Keywords:** Eigenvalues, Linearization, Lyapunov's method, Nonlinear systems, Non-linear systems, Perturbed linear autonomous, Stability properties.

## INTRODUCTION

A system, from an energy perspective, could be said to have stable equilibrium if it is in the least energy state or when its energy is non-increasing. It could be said to be stable if its solution evolved in such a way that a small change in the equilibrium point will not lead to a radical change in the behavior of the system. If it does, it is said to be unstable [1-4].

This chapter, the stability of linear perturbed systems including periodic ones is considered. Moreover, Lyapunov stability technique which is also called the "Lyapunov second method" is introduced and it is a generalization of the energy concept. The emphasis is on how to qualitatively investigate the stability properties of equilibrium of a system. We will consider a linearizing procedure and establish the stability properties of nonlinear systems from the linearized ones.

Furthermore, we will construct Lyapunov functions and use them to obtain some stability criteria for the equilibrium points of some systems. Different types of stability will be studied using the Lyapunov second method [5-7, 12].

### Stability of Linear Perturbed Systems

In this section, without loss of generality, the stability of perturbed linear autonomous systems will be closely studied, bearing in mind that the idea can be extended to non-autonomous systems.

**Benjamin Oyediran Oyelami**
**All rights reserved-© 2024 Bentham Science Publishers**

Stability of homogeneous differential equations is a prototype of stability theory [1] and sometimes such a differential equation may contain a perturbation function. We intend to study stability and its characterization for linear perturbed systems. Jack [11] provided a bundle of theorems for this purpose, especially for non-autonomous systems.

Consider the linear autonomous system:

$$\dot{x} = Ax + f(t,x) \tag{7.1}$$

where $A$ is a square matrix and $f(t,x)$ is the perturbation function, which is continuous, such that:

$$\|f(t,x)\| \leq m\|x\|,$$

for some positive constant $m$. We assert that the system is asymptotically stable [1, 12].

Using the above condition, it is not difficult to show that:

$$\|x\| \leq c\exp{-(\gamma - mc)(t - t_0)}\|x_0\|, \tag{7.2}$$

therefore, given $\eta > 0$, we can find $\sigma > 0$ such that $\|x_0\| \leq \sigma \Rightarrow \|x(t)\| < \eta$

*i.e.* $\|x\| < c\exp{-(\gamma - mc)(t - t_0)}\sigma < \eta$,

for $t > t_0 + T(\eta)$, where $T(\eta) = \dfrac{1}{\gamma - mc}\ln\left(\dfrac{\eta}{c\sigma}\right)$ for $\gamma - mc > 0$ and $\in > c\delta$. Uniform stability is implied from the definition. We note that $\|x(t)\| \to 0$ as $t \to \infty$ thus asymptotic stability follows and hence the proof.

**Remark 7.1**

The above result holds for non-autonomous systems (Jack [11], pp.87).

## Theorem 7.1

Let $\dot{x}(t) = A(t)x(t) + f(t, x(t))$ be a linear perturbed system where $A(t), f(t, x(t))$ satisfy conditions that allow solutions to exist and be uniquely determined in a given interval. Suppose also that $f(t, x(t))$ degenerates such that:

$$\|f(t, x(t))\| \leq \eta(t)\|x(t)\| \tag{7.3}$$

and $\int_{t_0}^{t} \eta(s)ds < m$ for some constant $m > 0$ and $t \geq t_0$.

Then the linear perturbed system is uniformly stable.

## Proof

The proof is by a variation of constant parameter and estimation as follows:

$$\|x(t)\| \leq e^{At}\|x_0\| + c\int_{t_0}^{t} \eta(s)\|x(s)\|ds$$

$$\|x(t)e^{\alpha t}\| \leq c\|e^{\alpha t}x_0\| + c\int_{t_0}^{t} \eta(s)\|x(s)e^{\alpha t}\|ds$$

Let $V(t) = x(t)e^{\alpha t}$, thus by Gronwall's inequality [1, 9, 10] we have

$$\|v(t)\| \leq c\|v_0\| + \int_{t_0}^{t} c\eta(s)\|v(s)\|ds$$

$$\leq c\|v_0\|\exp(\int_{t_0}^{t} c\eta(s))ds$$

That is:

$$\|x(t)\| \leq c\|x_0\|\exp-(\alpha - cm)(t - t_0) \tag{7.4}$$

The inequality in equation (7.4) is similar to the one in equation (7.2), but the system is not asymptotically stable. This can be justified by a counter-example. As an exercise construct a counter example.

**Stability of Linear Periodic Systems (LPS)**

Hill's equation is stable if the absolute value of the characteristic multipliers $p_1, p_2$ are one, *i.e.* $|p_1| = |p_2| = 1$ otherwise, it is unstable. This condition simply stipulates that the solutions are bounded, hence stable.

Suppose $x_1(t, \lambda)$ and $x_2(t, \lambda)$ are two linearly independent solutions of the Hill's equation, satisfying the conditions: $x_1(0, \lambda) = 0, \dot{x}_1(0, \lambda) = 1, x_2(0, \lambda) = 1, \dot{x}_2(0, \lambda) = 0$

Define $k = \frac{1}{2}(x_1 + \dot{x}_2)$. Then Hill's equation is stable if $|k| < 1$ and unstable if $|k| \geq 1$.

It will be recalled that Mathieu's equation is a particular case of Hill's equation for the case $\phi(t) = \cos 2t$. It follows that the Mathieu equation inherits the stability properties of Hill's equation [11].

Generally, for a periodic system, a necessary and sufficient condition for uniform stability is that the characteristic multipliers must have moduli less than one (All characteristic exponents have real parts equal to zero). For uniformly asymptotic stability, all the characteristic multipliers must have a modulus less than one (All characteristic exponents have negative real parts).

A reciprocal system ( $R^p$ ) is stable if all its solutions are bounded. This is, of course, another way of saying that the characteristic multipliers of the system must have an absolute value equal to one *i.e.*, $|p| = 1$ and none has a simple elementary divisor. This condition is necessary as well as sufficient too for stability for $R^P$.

In summary, for a periodic system and an autonomous system, the following chain of implications is true:

| Stability |
| :---: |

| Uniform stabilty |
| :---: |

| Asymptotic stability |
| :---: |

| Uniform asymptotic stability |
| :---: |

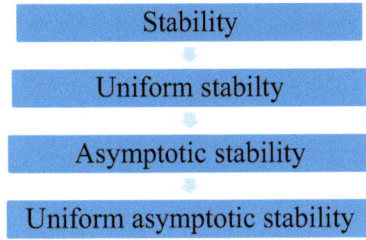

## Lyapunov Stability Theory

Lyapunov stability theory, also called "Lyapunov second method" is a generalization of the energy concept [12]. It could be said to have evolved from the fact that a system exists in stable equilibrium if it is at the least energy state or when its energy is non-increasing.

The theory of Lyapunov stability encompasses wider areas of applications and its advantage over other quantitative methods lies in the fact that it makes use of properties of scalar functional (The Lyapunov functional) to investigate stability instead of solving the problem explicitly [2,6,8].

It is worthy of note that the class of differential equations with closed solution forms is very small [3,4, and 5]. So, the Lyapunov stability theory becomes a useful and powerful tool to overcome shortcomings in using quantitative methods for investigating the stability of systems.

A very serious task confronting experts in the field of differential equations is the construction of the Lyapunov functional. Many techniques have evolved over the century on the subject matter.

Stephen [12] presented two elegant methods on this topic, while, on the other hand, Goh [2] and Jack [11] gave examples that practically demonstrated the construction of the Lyapunov functional of the form $v(x) = x^T B x$, where $B$ is a symmetric matrix satisfying an equation of the form:

$$A^T B + BA = -C \qquad (7.4)$$

where $A$ is the companion matrix of the system for which stability is to be investigated for some positive definite matrix $C$.

In Jack [11] (See Lemma 1.6), we are rest assured of the existence of $V(x)$ as long as the system under consideration is asymptotically stable, *i.e.*, the real part of the eigenvalues of $A$ are negative.

This method, though effective, but is time and energy sapping, especially for non-linear system. We will, therefore, resort to less cumbersome linearization process for non-linear systems. Consequently, we will study the behavior of the linearized system.

### Linearization

Suppose $g \in C^1(E \times E^n, E^n), g(t,0) = 0$ for all $t$ if $A(t)x = \left[\dfrac{\partial g(t,x)}{\partial x}\right]_{x=0} x$ and

$f(t,x) - A(t,x)$ then $f(t,0), \dfrac{\partial f(t,0)}{\partial x} = 0$ the function $A(t)x$ is linear approximate to the right-hand side of the equation:

$$\dot{x} = g(t,x) \ , \tag{7.5}$$

where $\left[\dfrac{\partial g(t,x)}{\partial x}\right]_{x=0}$ is the Jacobian matrix evaluated at: $x = 0, x = (x_1, x_2, ..., x_n) \in E^n$.

### Theorem 7.2

If all eigenvalues of $\left[\dfrac{\partial g(t,x)}{\partial x}\right]_{x=0}$ are distinct and have negative real part and

$$\lim_{\|x\| \to 0}\left[\frac{\|f(t,x) - A(t)x\|}{\|x\|}\right] \tag{7.6}$$

Then the critical (equilibrium) point $x = 0$ is asymptotically stable.

### Remark 7.2

If any of the four basic characterizations of the stability is satisfied by the Jacobian matrix $\left[\dfrac{\partial g(t,x)}{\partial x}\right]_{x=0}$ and in addition, the equation (7.6) is satisfied too, then the nonlinear system automatically inherits the same characterization of the stability as the Jacobian.

**Corollary 7.1**

Let $\dot{x} = g(t,x)$   where   $x = \begin{pmatrix} x \\ y \end{pmatrix} \in E^2, g(t,0) = 0$. Linearization of the given differential equation gives:

$$\dot{x} = ax + by + R_1(x,y)$$
$$\dot{y} = cx + dy + R_2(x,y)$$

(7.7)

and

$$\left[ \frac{\partial G(t,x)}{\partial x} \right]_{x=0} = \begin{vmatrix} \dfrac{\partial G_1}{\partial x} & \dfrac{\partial G_1}{\partial y} \\ \dfrac{\partial G_2}{\partial x} & \dfrac{\partial G_2}{\partial y} \end{vmatrix} = \begin{bmatrix} a & b \\ c & d \end{bmatrix}$$

Then trivial solution, $x = 0$ is asymptotically stable if $ad - cb > 0, a + d < 0$ and $\displaystyle\lim_{\|x\| \to 0} \frac{\|R(x)\|}{\|x\|} = 0$, where $x = (x_1, x_2)$.

**Proof**

$\dot{x} = Ax + R(x)$ has the characteristic equation $\Im(\lambda) = \lambda^2 - (a+b)\lambda + ad - bc = 0$. The necessary condition for asymptotic stability of $x = 0$ is that the coefficients of the roots of $\Im(\lambda)$ are positive, by the Routh- Hurwitz criterion; that is, $ad - bc > 0, a + d < 0$ and the above is therefore established by Theorem 7.1, if $\displaystyle\lim_{\|x\| \to 0} \frac{\|R(x)\|}{\|x\|} = 0$.

Let consider the autonomous system:

$$\dot{x} = f(x)$$

(7.8)

Where $f \in C^0(E^n, E^n)$ since our investigation will be restricted to the stability of trivial solution. Let us limit ourselves to $D$ (open subsets of $E^n$).

Define $D = \{x \in E^n : \|x\| < H, H > 0\}$ rather than the whole Euclidean space $E^n$. Assuming $f : D \to E^n$ is a continuous function such that the solution of the equation (13.8) is uniquely determined for a given finite initial data, $t_0$ $|t_0| < \infty, x_0 \in D, \|x_0\| < D$.

## Definition 7.1

Let $D \subset E^n$ be an open set (as defined above) in $E^n$ with $0 \in D$. A scalar function $V(x) > 0, V(0) = 0$ for $x \neq 0$ is said to be positive definite; $V(x)$ is negative semi definite (negative definite) in $D$ if $-V(x)$ is semi definite (positive definite) in $D$.

## Definition 7.2

We define a Lyapunov function $V(x)$ as follows:

1. $V(x)$ and all its partial derivative $\dfrac{\partial V}{\partial x}$ the continuous, $j \in \{1, 2, 3, ..., n\}$.

2. $V(x)$ is positive definite $i.e. V(0) = 0, V(x) > 0$ for $x \neq 0$.

The derivative of $V(x)$ along the solution path, is define as:

$$V(x) = \sum_{i=1}^{n} \frac{\partial V(x_i)}{\partial x_i} x_i = \sum_{i=1}^{n} \frac{\partial V(x_i)}{\partial x_i} f(x_i) \tag{7.9}$$

The equation (7.9) is called the Euler derivative. Suppose $\dot{V}(x)$ is negative definite such that $\dot{V}(0) = 0, \forall x \in D$ along the solution path of (7.1) satisfying enough smoothness condition to ensure unique solution exists. Then $V(x)$ is called a Lyapunov function [5-7].

## Example 7.1

Consider the following system of ordinary differential equation:

$$\dot{x} = x(1 - 2x^2 - y^2) + y$$

$$\dot{y} = y(1 - x^2 - 2y^2) - x$$

Show that the origin is an unstable equilibrium point and that every solution is unbounded.

**Solution**

$$\text{Jacobian matrix } A = \left[\frac{\partial G(t,x)}{\partial x}\right]_{x=0} = \begin{vmatrix} \dfrac{\partial G_1}{\partial x} & \dfrac{\partial G_1}{\partial y} \\ \dfrac{\partial G_2}{\partial x} & \dfrac{\partial G_2}{\partial y} \end{vmatrix}_{(0,0)} = \begin{bmatrix} 1 & 0 \\ 0 & 1 \end{bmatrix}$$

Using the equation (7.7) we have:

$$\dot{x} = x + R_1(x, y)$$

$$\dot{y} = y + R_1(x, y)$$

The Eigenvalues of $A$ are $\lambda_1 = \lambda_2 = 1$.

$$R(x, y) = \begin{pmatrix} R_1(x, y) \\ R_2(x, y) \end{pmatrix} = F(t, x) - Ax$$

$$= \begin{pmatrix} x - 2x^3 - xy^2 + y \\ y - x^2y - 2y^2 - x \end{pmatrix} - \begin{pmatrix} 1 & 0 \\ 0 & 1 \end{pmatrix}$$

$$= \begin{pmatrix} -2x^3 - xy^2 + y \\ -x^2y - 2y^3 - x \end{pmatrix}$$

$$|R(x, y)| = \left|-2x^3 - xy^2 + y\right| + \left|-x^2y - 2y^3 - x\right|$$

$$\leq 2\left(|x|^3 + |y|^3\right) + |xy|\left(|x| + |y|\right) + \left(|x| + |y|\right)$$

$$\leq 2\left(|x|^3 + |y|^3\right) + |xy|\left(|x| + |y|\right)^3 + \left(|x| + |y|\right)^3$$

$$= 4\left(|x| + |y|\right)^3 = 4\|x\|^3$$

Since $$\|x=(x_1,x_2)\|=|x_1|+|x_2|.$$

$$\lim_{\|x\|\to 0}\frac{\|R(x,y)\|}{\|(x,y)\|}=\lim_{\|x\|\to 0}\frac{\|4(|x|+|y|)\|}{|x|+|y|}=4\|x\|^2$$

Hence by Theorem 7.1, the equilibrium point is not stable and unstable solution cannot be bounded.

## Example 7.2

Show that the trivial solution of the following equation is stable:

$$\dddot{x}+2(1-\dot{x}^2)\dot{x}+x=0$$

## Solution

Let
$$\dot{x}=x_1$$

$$\dot{x}_1=-2(1-x_1^2)x_1-x$$

$$A=\left[\frac{\partial G(t,x)}{\partial x}\right]_{x=0}=\begin{vmatrix}\dfrac{\partial G_1}{\partial x}&\dfrac{\partial G_1}{\partial y}\\[2mm]\dfrac{\partial G_2}{\partial x}&\dfrac{\partial G_2}{\partial y}\end{vmatrix}_{(0,0)}=\begin{bmatrix}0&-1\\1&-2\end{bmatrix}$$

$$|A-\lambda I|=\begin{vmatrix}-\lambda&-1\\-1-\lambda&-2-\lambda\end{vmatrix}=(-\lambda-1)\begin{vmatrix}1&-1\\1&-2-\lambda\end{vmatrix}$$

$$=(-\lambda-1)(-2-\lambda+1)=(1+\lambda)^2$$

$$\lambda_1=\lambda_2=-1.$$

$$R(x,y)=F(t,x)-Ax=\begin{pmatrix}x_1\\-2x_1+2x_1^2-x\end{pmatrix}-\begin{pmatrix}0&-1\\1&-2\end{pmatrix}\begin{pmatrix}x\\x_1\end{pmatrix}$$

$$=\begin{pmatrix}2x_1\\2x_1^3-2x\end{pmatrix}$$

$$\frac{\|R(x_1,x)\|}{\|(x_1,x)\|} \le \frac{2(|x_1|+|x_1|^3+2|x|)}{|x_1|+|x|}$$

$$\le \frac{3(|x|+|x_1|)^3}{|x_1|+|x|} = \frac{3\|x\|^3}{\|x\|} = 3\|x\|^2$$

$$\lim_{\|x\|\to 0} \frac{\|R(x,y)\|}{\|(x,y)\|} = 0$$

Thus, by Theorem 7.1, the Proof is justified.

**Problem 7.1**

1. Giving the equation:

$$\dddot{x} + f(x)\,\ddot{x} + 3\dot{x} + 6x = 0$$

$f$ is a real-valued continuous function such that for some positive constant, $c$, $f(y) \ge c > 2$, by considering the function

$$V(x,y,z) = \frac{3}{2}z^2 + 3yz + 3\int_0^y f(u)u\,du + \frac{1}{2}(3x+2y)^2$$

or otherwise, show that the trivial solution of $\dot{x} = f(x), f(0) = 0$ is stable where $f \in C^0(\Omega, E^n), \Omega = \{x \in E : \|x\| < k, k > 0\}$ is stable.

2. Show that the trivial solution of $\ddot{x} + a\dot{x} + bx = 0$ is asymptotically stable if $a > 0$ and unstable if $a < 0$.

3. Consider the system:

$$\dot{x}_1 = -x_1 + x_2^2$$

$$\dot{x}_2 = x_1^2 - 2x_2$$

Prove that the trivial solution $x = 0$ is uniformly asymptotical stable.

4. Let $\ddot{x}(t) + \alpha x(t) + h(t) = \cos t$, where $\alpha > 0, \dfrac{h(x)}{x} > 0$ for all $x$. Using

$$V(x,y) = \frac{1}{2}(y+\alpha x)^2 + \frac{1}{2}y^2 + 2\int_0^x h(s)\,ds$$

show that the trivial solution is stable and hence there exists a periodic solution of the given differential equation.

5. Let $f(x,y) \in C^1(E^2, E)$ such that $f(0,0) = 0$, $f(x,y) > 0$ if $(x,y) \neq 0$. Investigate the stability of the following system:

$$\dot{x} = y - xf(x,y)$$

$$\dot{y} = -x - yf(x,y)$$

Using $V(x,y) = k\left(x^2 + y^2\right), k > 0$

Show also that the trivial solution of $\dot{x}_1 = -x_1 + x_2^2$ is uniformly asymptotically stable.

$$\dot{x}_2 = x_1^2 - 2x_2$$

6. Consider the system:

$\dot{x}(t) = A(t)x + f(t,x) + b(t), f \in C^0(E^n)$ and for each $\in > 0$, suppose that given $\in > 0$, there exists $\sigma > 0$ such that $\|f(t,x)\| < \in \|x\|$ for $\|x\| < \sigma, t \in E$. Prove that the system $\dot{x}(t) = A(t)x(t)$ is uniformly asymptotically stable for $b(t) \in C^0(E^n), \lim_{t \to \infty} b(t) = 0$.

7. Show that the autonomous system:

$$\dot{x} = x - y$$

$$\dot{y} = 4x^2 + 2y^2 - 2$$

has the critical (equilibrium) points at $(1,1)$ and $(-1,-1)$ and that both of them are unstable.

8. Show that the autonomous system:

$$\dot{x} = \frac{1}{2}e^{2x} \sin y + 2\sin x \cos x + e^z$$

$$\dot{y} = \sin(2x + 2y)$$

$$\dot{z} = \tan(2x + 2)$$

has unstable critical point at $x = y = z = 0$

9. Investigate the stability of the system:

$$\dot{x}_1 = 2x_1 + x_2 - 5x_2^2$$

$$\dot{x}_2 = 3x_1 + x_2 + \frac{1}{2}x_2^2$$

10. Investigate the stability of the system:

$$\dot{x}_1 = 2x_1 + 8\sin x_2$$

$$\dot{x}_2 = 2 - e^{x_1} - 3x_2 - \cos x_2$$

## CONCLUSION

This chapter, the stability property of linear perturbed systems including the periodic ones is considered. The Lyapunov second method is introduced to qualitatively investigate stability properties of equilibrium of a system. A linearizing procedure is presented and used to investigate the stability of nonlinear systems using the linearized ones. Lyapunov functions are constructed and used to obtain some criteria for stability of equilibrium points of some systems.

## REFERENCES

[1]    G. Birkhoff, and Rota Gian-Carlo, *Ordinary Differential Equations..* Wiley, 1989.

[2]    B.S. Goh, *Stability of some Multispecies Population Models in Modeling and Differential Equations in Biology..* Lecture Notes in Pure and Applied Mathematics Marcel Dekker, Inc. Publication: New York, 1980, pp. 209-216.

[3]    Stephen. Grossberg, *Studies of Mind and Brain: Neural Principles of Learning, Perception, Development, Cognition, and Motor Control..* Reidel Press, 1987.

[4]    P.V. Gupta, and P.C. Dhar, *Network Analysis and Synthesis.* 'Dhan Pat Raj Publication: Delhi, India, 2006.

[5]    J.J. Hopfield, "Neurons, dynamics and computation", *Phys. Today,* vol. 47, no. 2, pp. 40-46, 1994.

       http://dx.doi.org/10.1063/1.881412

[6]    M. Martcheva, "Avian Flu: Modeling and implications for control", *J. Biol. Syst.,* vol. 22, no. 1, pp. 151-175, 2014.

       http://dx.doi.org/10.1142/S0218339014500090

[7]    R.M. May, *Stability.* Princeton University Press: Princeton, NJ, 1974.

[8]    V.G. Matsenko, and V.N. Rubainovski, ""Application of lyapunov direct method for analyzing the age structure of biological population", (Russian)", *Zhvyscist Math. Fiz,* vol. 23, no. 2, pp. 320-332, 1983.

[9]    B.O. Oyelami, S.O. Ale, and M.S. Sesay, "On existence", *Abuja Conference in Ordinary Differential Equations.,* 2000, pp. 101-117.

[10]     B.O. Oyelami, S.O. Ale, and M.S. Sesay, "Impulsive cone value integrodifferential and differential inequalities", *Electron. J. Differ. Equ.,* vol. 66, 2005pp. 1-14.

[11]     K. Hale Jack, *Ordinary Differential Equations..* Wiley-Interscience: New York, London, Sydney, Toronto, 1969.

[12]     S. Barnett, *Cameron..* Clarendon Press: USA, 1985.

# CHAPTER 8

# Stability Property of Some Neural Firing and Avian Influenza Infection Models

**Abstract:** The stability property of equilibrium points of some neural network models is investigated. We have introduced different types of Lyapunov functions to carry out the investigation. The models considered are: Grossberg, Hopfield, Fitz-Nagomo and Fitzhugh models, respectively. The equilibrium points and stability conditions are obtained for the Avian influenza infection. The conditions for bio economic equilibrium points for the fish model were also obtained.

**Keywords:** Bio-economic equilibrium point, Equilibrium points, Fish model, Fitz-Nagomo models, Fitzhugh model, Grossberg, Hopfield, Lyapunov functions, Neural network models, Stability property.

## INTRODUCTION

Neural network is primarily concerned with modelling the activity of the brain, its behavioral processes, and the application of these models to computers and related technologies [1,4,6,7,10]. The Areas where neural network find useful applications are neuroscience, artificial intelligence, vision and image processing, speech and language understanding, pattern recognition, parallel distributed processing, and so on (Gene *et al*. [4]; Hopfield [8]).

An Artificial neural network (ANN) is an information or signal processing entity containing elements, called artificial neurons, or sometimes referred to as nodes. The neurons are interconnected by direct links called connectives, which perform parallel distributed processing (PDP) in order to solve the desired biological or computational task [5,7,9, 20].

Artificial neural networks can be used effectively to solve many scientific and engineering problems that are formulated mainly as variational or optimization problems derived from the learning equation with or without a teacher [4].

This is a richly connected network of simple computational elements modelled as biological processes. Their origin can be traced back to the late nineteenth and early twentieth centuries when psychologists tried to identify the neural basis of intelligence [4].

**Benjamin Oyediran Oyelami**
**All rights reserved-© 2024 Bentham Science Publishers**

MacCulloch initiated research on the central nervous system in 1943 and a neural network model which was published with Walter Pitts. In 1949, Donald Hebb proposed a model for learning in neural network and Dean Endermonds in 1951 proposed a similar model but on electromechanical learning machine which incorporated the ideas in a motor-driven memory with forty control knobs Cichocki and Unbehaven [1].

The discipline Artificial Intelligence was introduced in 1956 at the Dartmouth Conference where Anderson James presented a paper based upon his development on brain state in a box (BSB) (Nicholas [9]).

Grossberg developed a mathematical model in 1982, which encompassed a variety of neural network models as well as population biology and macro molecular evolution.

In this chapter, we consider four models of neural network types, namely; Grossberg, Hopfield, Fitzhugh-Nagumo and Fitzhugh models. Equilibrium points of the models are determined and consequently stability properties of systems investigated using a series of Lyapunov functions which we introduced. The fundamental problem we encountered was how to obtain suitable Lyapunov functions for the models. The reason why we opted for the stability of the models is that it offers a precondition for optimization of the models [11-15].

## Preliminary Definitions

### Neuron

This is the basic unit of the central nervous system (CNS) that sends signals through the neural network.

### Synapse

This is the connection junction between two neurons.

Neural networks are a class of models inspired by the neural circuitry in human and animal brains. They occupy a spectrum, ranging from <u>artificial neural networks</u> (ANNs) to <u>biological neural networks</u> (BNNs).

## MACCULLOCH -PITT Model

The first real model of a nerve cell that could be simulated on a computer was developed by McCulloch and Pitts (Fig. **8.1**).

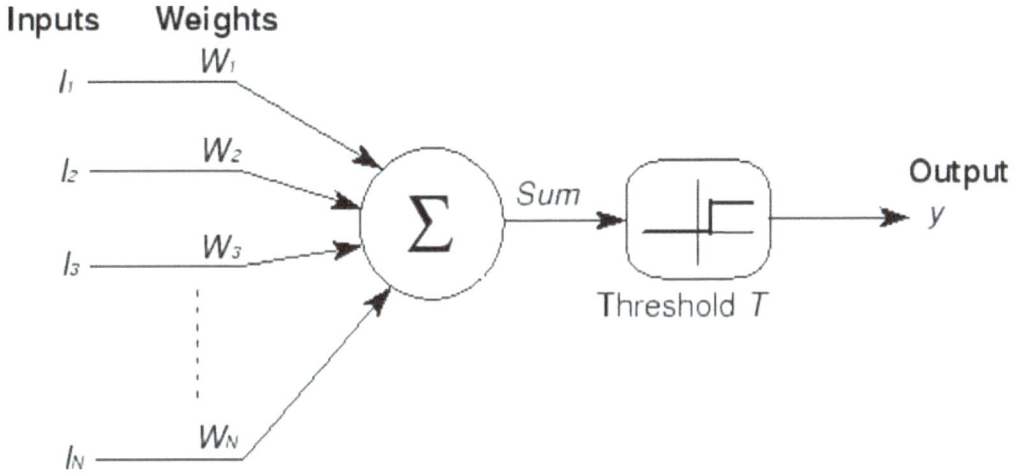

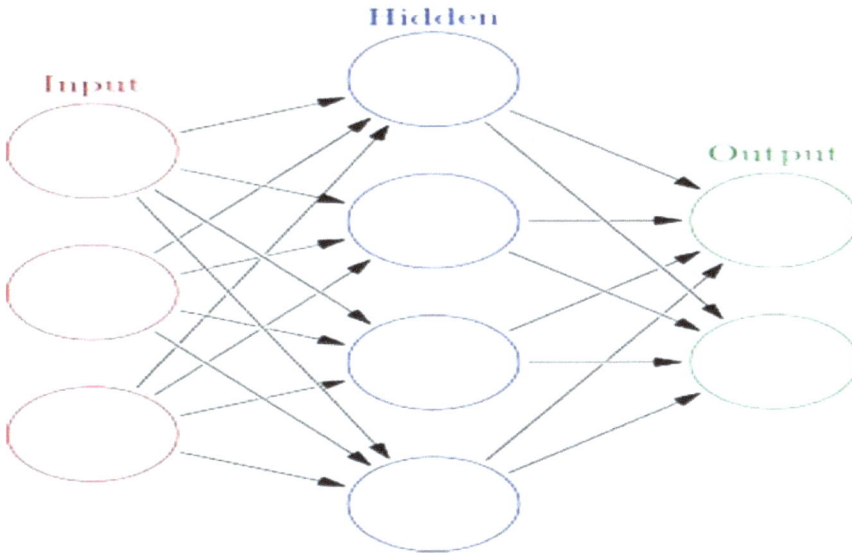

*(Fig. 8.1) contd.....*

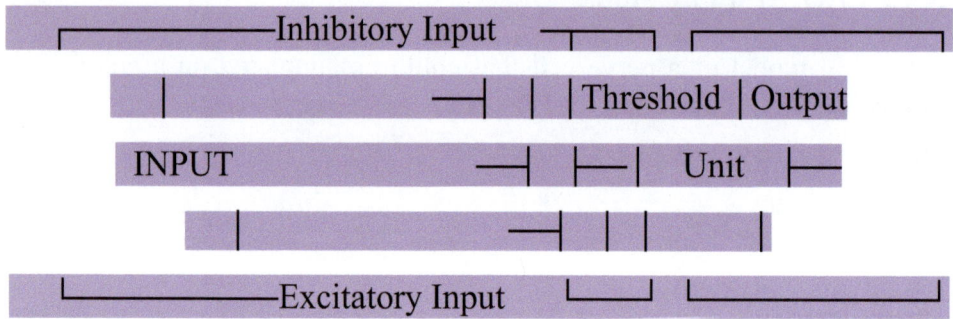

**Fig. (8.1).** McCulloch-Pitt Model.
Source: from Wikipedia

The human neutron system contained several networks of interconnecting neurons; each performs a discrete computation at any given moment. The results of this computation are transmitted to another neuron along the pathway.

A single neuron sends the results to as many as 10,000 other neurons as signal to the input of the neuron in the form of voltage. These signals either inhibit the other neurons from sending signal in or excite them to send signals to another neuron along the pathway.

**A Neuron Model**

Fig. (**8.2**) is diagram of a neuron anatomy. Neuron is the basic unit neural network:

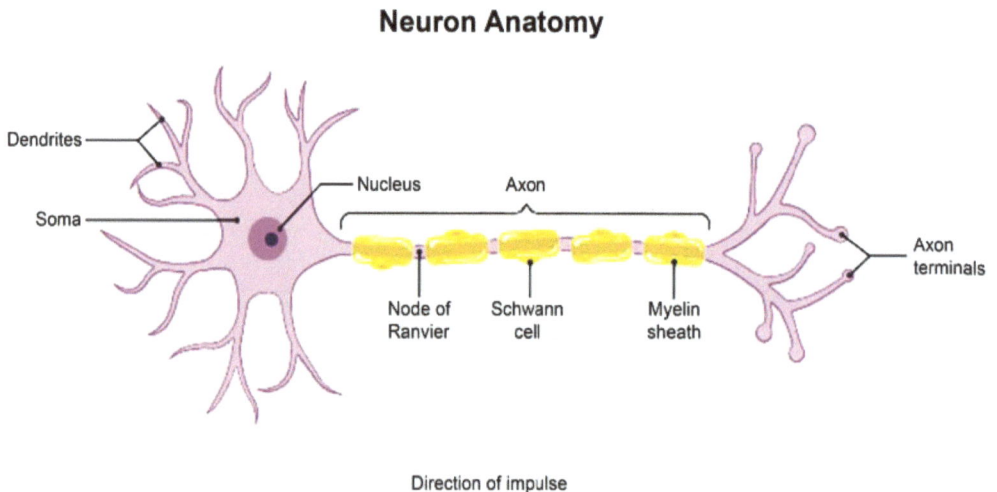

**Fig. (8.2).** A Neuron anatomy.

The adjustable multiplicative weight corresponds to biological synapses. For the purpose of analytical modelling, it is often convenient to allow a positive weight to represent excitatory connection and/or negative weight for an inhibitory connection. A weigh of zero is assigned when no connection between a pair of neurons is to be made.

## Application of Lyapunov method to the neural models

We will consider the differential equation which is an underlying equation for neural models.

This equation is as follows:

$$\dot{x}_i = f(t, x_i), t \in [t_0, +\infty) = E^1 \tag{8.1}$$

$$f : E^+ \times E^n \to E^n ,$$

where $f$ is continuous and locally Lipschitzian with respect to $x_i$. This allows the existence of the solution of equation (8.1) passing through $(t_0, x_0) \in E^+ \times E^n$. Corresponding to equation (8.1) is the following perturbed equation [3]:

$$\dot{x}_i = f(t, x_i) + g(t, x_i), t \in [t_0, +\infty) = E^1 . \tag{8.2}$$

Where,

$$f, g : E^+ \times E^n \to E^n$$

$f$ and $g$ satisfy the regular conditions for the existence of solutions of equation (8.2). To investigate the stability of the neural model, we need the function

$$V : (t_0, \infty) \times E^n \to E ,$$

which is continuous and whose Dini derivative [11] along the solution path of equation (8.1) is defined by:

$$\dot{V}_{(15.1)}(t, x) = \lim \frac{1}{h} [V(t + h, x + hf) - V(t, x)] \tag{8.3}$$

$V$ will be assumed to satisfy certain conditions.

The following theorems will be useful for obtaining our results:

**Theorem 8.1** (Jack [11], Theorem 4.2, pp309)

Suppose $f$ is such that $f(t,0) = 0$   $f(t,x)$ is locally Lipschitzian in $x$ uniformly with respect to $t$ and the solution $x = 0$ is uniformly asymptotically stable. Then there exist positive constants $r_1 > 0$ and $k = k(r_1) > 0$, a positive definite function $b(r)$, a positive function $C(r)$ on $0 \le r \le r_1$ and a scalar function $V(t,x)$ defined and continuous for $t > 0, x \in E^n$, $\|x\| \le r_1$ such that:

a.                 $a(\|x\|) \le V(t,x) \le b(\|x\|)$

b.                 $V_{(1)}(t,x) \le - C(\|x\|)V(t,x) \le - \|x\|C(\|x\|)$

c.                 $\|V(t,x) - V(t,y)\| \le k\|x - y\|$

for all $t \ge 0, x, y \in E^n, \|x\| \le r_1, \|y\| \le r_1$.

The above theorem shows that if $x = 0$ is the trivial solution of equation (8.1) which is stable, then the Lyapunov function $V(t,x)$ can also be found to satisfy conditions (a – b).

**Theorem 8.2** (Jack [11], Lemma 5.1, pp311)

Suppose $f$ satisfies the condition in Theorem 8.1and $V$ is the function given in that theorem, $g(t,x)$ is any continuous function on $E^+ \times E^n$ into $E^m$. Consider the equation:

$$\tilde{x} = f(t, x) + g(t, x) \tag{8.3}$$

Then,

$$\tilde{V}_{(2)}(t, x) \le - C(\| x \|)V(t, x) + K \mid g(t, x) \mid \tag{8.4}$$

for all $t \ge 0, \|x\| \le C_1$.

## Remark 8.1

If there exist $\delta > 0$ and T > 0, such that $\|g(t,x)\| < \delta$ for any $x_0 \varepsilon E^n$, then the solution $x(t) = x(t,t_0,x_0)$ of equation (8.3) satisfying $x(t_0) = x_0$ is uniformly stable. Thus, to obtain results on stability, we obtain the results by application of Theorem 8. 1 and Theorem 8. 2.

## Grossberg model

Grossberg model as was emphasized in the introduction was developed in 1982. The analysis of this model made by Cohen and associates in 1983, yielded conditions which the system of the differential equations used to characterize a number of popular neural network models which converge to stable equilibrium states.

The model which they analyzed is a dynamical system of mutually interdependent differential equation of the form (general purpose learning equation):

$$\frac{dx_i}{dt} = a_i(x_i)\left[ b_i(x_i) - \sum_{j=1}^{N} c_{ij} g_i(x_i) \right]$$

$$(8.6)$$

where

$a_i(x_i)$ is the neural output related to internal activity by the equation $x_i = a(x_i)$

$b_i(x_i)$ is the neural output as it jumps from one synapse to another

$x_i$ is the activity of the neuron

$c_{ij}$ is the activity of the synapse being in the form of excitatory and inhibitory synapse

$g_i(x_i)$ neural output.

Cohen and associates in 1987 (see Cichocki and Unbehuen, 1993) obtained a series of results that ensure that the solution of Grossberg model converges to the equilibrium points, that is, the set of internal activities of the neuron is such that:

$$f(x_i) = a(x_i) \left[ b(x_i) - \sum_{j=1}^{N} c_{ij} g_i(x_i) \right] = 0$$

$$(8.6)$$

We investigate the perturbed form of Grossberg model in the form:

$$\frac{dx_i}{dt} = a_i(x_i)[b_i(x_i) - \sum_{j=1}^{N} c_{ij} g_i(x_i)] + h(t, x_i)$$

$$(8.7)$$

such that $a_i(0) = 0, b_i(0) = 0, g_i(0) = 0$. Where a non-linear perturbation function $h(t,x)$ is introduced.

We can show that if the solution of equation (8.5) exists, we can construct a Lyapunov function such that the following conditions are satisfied:

The matrix $[c_{ij}]$ and the functions $a_i(x), b_i(x)$ and $g_i(x)$ are continuous such that:

$$f(t, x_i) = a_i(x_i) \left[ b_i(x_i) + \sum_{j=1}^{n} c_{ij} g_i(x_i) \right] + h(t, x_i)$$

$$(8.8)$$

And such that:

$$1. \max_{j} | a_i(x_i) b_i(x_i) - a_i(x_j) b_i(x_j) | \leq k_1 | x_i - x_j |$$

$$2. \max_{l,j} | a_i(x_i) c_{ij} - c(x_i) c_{ij} | \leq L_{ij}^* | x_i - x_j |$$

and

$$| h(t, x_i) - h(t, x_j) | \leq k_2 | x_i - x_j |$$

Then, it is easy to see that:

$$| f(t, x_i) - f(t, x_j) | \leq (k_1 + L_1 L_{ij}^* + k_2) | x_i - x_j |$$

Also, $f(t,0) = 0$. Thus, Theorem 8. 2 guarantees that the solution is uniformly asymptotically stable, provided $h(t, x_i) < \delta$, where $\delta$ can be found. There is a Lyapunov function that satisfies condition (1-3). The question is how do we construct such a Lyapunov function? Construct of Lyapunov functions, generally is very difficult, but, we consider this Lyapunov function:

$$V(t, x_i) = X_i^T A X_i , \tag{8.9}$$

where A is a positive definite i by i matrix. Differentiating equation (8.9), we set

$$\dot{V}(t, x_i) = \dot{x}_i^T A x_i + x_i^{TAx_i}$$

$$= a_i(x_i) \left( b_i(x_i) - \sum_{j=1}^{N} c_{ij} g_i(x_i) \right)^T A x_i + \left( a_i(x) h_i(t, x_i) \right)^T A x_i$$

$$+ x_i^T A \left( a_i(x_i) \left( b_i(x_i) - \sum_{j=1}^{N} c_{ij} g_i(x_i) + h(t, x_i) \right) \right)$$

Let:

$$b_i a(x_i) A - \Sigma g_i^T(x) c_{ij} A + A sup T h^T(t, x_i) = x_i^T Q \tag{8.10}$$

And:

$$A a_i(x_i) b_i(x_i) - A a_i(x_i) \Sigma c_{ij} g_i(x_i) + h_i(t, x_i) A = W x_i \tag{8.11}$$

Clearly, $\dot{V}(t, x_i) = x_i^T (W + Q) x_i \leq - \varepsilon x_i^T A x_i = - \varepsilon v(t, x_i) + k | h(t, x) |$,

where $k \geq \max [1, | h(t, x_i) |]$

This result guarantees that the trivial solution $x_i = 0$ of the modified Grossberg model with non-linear perturbation function $h(t, x_i)$ is uniformly asymptotically stable.

**Hopfield Continuous Model**

The behavior of a neuron is characterized by its activation level which is governed by the differential equation:

$$\frac{du_i}{dt} = -\frac{u_i}{\eta} + \Sigma T_{ij} g_i(u_i) + I_i$$

(8.12)

$\dfrac{u_i}{\eta_i}$ is a passive decay term.

$T_{ij}$ is the strength of the interconnection between neuron j and neuron i, $g_i(u_i)$ is the activation function of neuron i and $I_i$ is the external input to neuron i.

The activation level $u_i$ is a continuous variable potential in the biological neurons.

In 1984, Hopfield discovered a Lyapunov function for a neural network of n-neuron characterized by equation (8.12) above which can be expressed as:

$$V(t, x_i) = -\frac{1}{2} \sum_{\ell=1}^{N} \sum_{j=1}^{N} T_{ij} g_i(u_i) - \Sigma I_i g_i(u_i)$$

Where the gain of the activation function is sufficiently high enough (*e.g.* take cardinality of the continuous). We shall use the following condition to investigate the stability of the model:

Suppose that,

A1. $T_{ij}$ is a positive definite matrix

A2.
$$\frac{\partial g_i(u_i)}{\partial u_i} \frac{u_i}{\eta} \geq g_i(u_i)$$

A3. There exist positive constants $\delta_1$, $\delta_2$ and $\delta$ such that

$$\|g_i(u_i)\| < \delta_i, \sum_{i,j}^{n} |T_{ij}| < \delta_2$$

Then by differentiating $V(t, x_i)$ with respect to $t$, we get:

$$-\frac{dV(t, u_i)}{dt} = \sum_{i=1}^{N} T_{ij} \frac{\partial g_i(u_i)}{\partial u_i} \cdot \frac{u_i}{\eta} + \Sigma \Sigma T_{ij} \frac{\partial g_i(u_i)}{\partial u_i}$$

$$+ \sum_j T_{ij} g_i(u_i) + \sum_i \sum_j b_{ij} T_{ij} + \sum_i T_i \frac{\partial g_i(u_i)}{\partial u_i} \cdot \frac{u_i}{\eta}$$

$$- \sum_j T_i \frac{\partial g_i(u_i)}{\partial u_i} \Sigma T_{ij} g_i(u_i) - \Sigma I_i^2 ]$$

$$\frac{dV(t, u_i)}{dt} \le -\frac{1}{2} \sum_i \sum_j T_{ij} g_i(u_i) - \sum_i \sum_j T_{ij} g_i(u) \frac{\eta}{u_i}$$

$$+ \Sigma T_{ij} g_i(u_i) + \Sigma T_{ij}^2 - \Sigma I_i g_i(u_i) \le V(t, u_i) + g(t, u_i)$$

$$g(t, u_i) = \Sigma T_{ij} g_i(u_i) + \Sigma T_{ij}^2 - \Sigma \Sigma T_{ij} g_i(u) \frac{\eta}{u_i}$$

Clearly, from A3 we have:

$$|g(t, u_i)| \le \left| \delta_1 \left( 1 + |\frac{\eta}{u_i}| \right) \delta_2 + \delta_1 \right| < \delta$$

The result from Theorem 8. 2 shows that the equilibrium point of Hopfield model is uniformly stable using the perturbed function $g_i(u_i)$ in A3.

## Fitzhugh-Nagumo Model of Neuron Firing

Fitzhugh-Nagumo model [17] is a simplification of the model due to Hodgkin and Huxley. The model relates the potential of the cell membrane, the permeability of the membrane and the applied current. This model is given by the equation:

$$\tilde{V} = f(v) - \omega + I$$

$$\dot{\omega} = bv - \gamma\omega$$

Where $f(v)=v(a-v)(v-I)$ represents the membrane potential $\omega$, the restoring force and I is the applied current, $a, b$ and $\gamma$ are constants, and $\omega$ is the restoring force.

The equilibrium point for model can be found from:

$$f(v) - \omega + I = 0 \qquad\qquad\qquad (8.13)$$

$$bv - \gamma\omega = 0 \qquad\qquad\qquad (8.14)$$

Combining equations(8.13) & (8.14), we have the following equilibrium equation:

$$c\omega^3 + d\omega^2 + e\omega + h = 0 \qquad\qquad\qquad (8.15)$$

$$v = \frac{\gamma}{b}\omega \qquad\qquad\qquad (8.16)$$

Where $\qquad c = \gamma^3, d = -(1+a)b\gamma^2, e = a\gamma b^2 + b^2 = (1+a\gamma)b^2, h = -b^3 I$ .

The analytic solution of equations (8.15) & (8.16), which is the equilibrium of the Fitzhugh-Nagumo's neuron firing model, is not easily obtainable in concrete terms. We may use the following iterative equations to obtain an approximate solution to it.

$$\alpha_{k+1} = -\frac{(d + 2\gamma_k) \pm \sqrt{d^2 + 4\gamma_k^2}}{2}$$

$$\beta_{k+1} = 2(d + \gamma_k) \pm \sqrt{d^2 + 4\gamma_k^2}$$

$$\gamma_{k+1} = \frac{f}{-\left(d + 2\gamma_k\right) \pm \sqrt{d^2 + 4\gamma_k^2}\left(2(d + 4\gamma_k) + \sqrt{d^2 + 4\gamma_k^2}\right)}$$

From above recursive relation, the value for the equilibrium can be generated once the starting points $\alpha_0, \beta_0$ and $\gamma_0$ are known and so as the number of iterations increases, the iteration will converge to the points $\alpha^*, \beta^*$ and $\gamma^*$, which are the equilibrium points for stable firing. For us to investigate the stability of the model, we require a Lyapunov function with positivity property and whose Dini derivative along the solution path has nice negativity property [21] & [22]. For this purpose, let us try $V(t, V, \omega)$ with positive definite property such that:

$$V(t, v, w) = \sum_{k=1}^{n} \sum_{l=1}^{n} V_{(\omega)}(t, v_k, \omega) \tag{8.17}$$

Thus:

$$\frac{dV}{dt} = \sum_{i=1}^{\infty} \sum_{j=1}^{\infty} \left\{ a_k^{(i)} V_k \dot{V}_k + b_k^{(i)} \omega_k \dot{\omega}_k \right\} + \frac{d f_{(i)}}{dt}$$

Let us choose

$$a_k^{(i)} f_k - a_k^{(i)} \omega_k + a_k^{(i)} + b_k^{(i)^2} \omega_k \le -\delta I$$

*i.e.*, choose $\omega_k$ such that:

$$\omega_k \le -\frac{\delta I + a_k^{(i)}(1 + f_k)}{b_k^{(i)2} - a_k^{(i)}} = -\left( \frac{\delta I + a_k^{(i)}(1 + f_k)}{b_k^{(i)2} - a_k^{(i)}} \right) \tag{8.18}$$

But:

$$\frac{df}{dt} = \frac{\partial f}{\partial V_k} \cdot \frac{dV_k}{dt} + \frac{\partial f}{\partial \omega_k} \frac{d\omega_k}{dt} \tag{8.19}$$

$$= (f(v_k) - \omega_k + I)\frac{\partial f}{\partial v_k} (bv_k - \gamma\omega_k)\frac{\partial f}{\partial \omega_k} = g(t, x).$$

Where $x = (v_k, \omega_k)$

*i.e.*

$$\frac{\partial f}{\partial t} = g(t, x) \tag{8.20}$$

Then by selecting $\delta = \max_k [\delta_{1k}^*, \delta_{2k}^*]$

Where:

$$\delta_{1k}^* = \max_k \left[ |f(V_k) - \omega_k + I|, |bV_k - \gamma\,\omega_k| \right] \tag{8.21}$$

$$\delta_{2k}^* = \max_k \left[ |\frac{\partial f}{\partial V_k}|, |\frac{\partial f}{\partial \omega_k}| \right] \tag{8.22}$$

It follows that $|g(t,x)| < \delta$. Hence by Theorem 8.2, the equilibrium points of Fitzhugh-Nagumo's model entertains uniform asymptotic stability.

**Fitzhugh Model**

We consider the Fitzhugh model as a Van der Pol model with some term added (inducing asymmetry). It also includes a model of arrhythmia:

$$\frac{dx_i}{dt} = y_i - 1.5 + \left( x_i - \left( \frac{x_i^3}{3} \right) \right) - A(\sin(0.2t)) \tag{8.23}$$

$$\frac{dy_i}{dt} = -\left( \frac{1}{9} \right)(x_i - 0.467 + 0.8(y_i - 1.5)) \tag{8.24}$$

We try:

$V(t, x, y) = ax^2 + by^2, a, b > 0$ real numbers.

The equilibrium points for the Fitzhugh model can be obtained from:

$$\frac{0.8}{3} x^3 + 0.2x - 0.467 + 0.8A\sin 0.2t = 0 \tag{8.25}$$

$$y + \frac{1}{0.8} x - 1.5 - \frac{0.467}{0.8} = 0 \tag{8.26}$$

From Newton- Raphson method, we can obtain approximate solutions to equilibrium equation as follows:

$$\text{Let } f_1(x_i) = \frac{0.8}{3} x^3 + 0.2x - 0.467 + 0.8A\sin 0.2t \tag{8.26}$$

$$f_2(x_2) = y - \frac{1}{0.8} x - 1.5 - \frac{0.467}{0.8} \tag{8.27}$$

We define:

$$x_{k+1} = x_k - \frac{f_1(x_k)}{f_{1'}^{\,'}(x_k)}, \quad f'(x_k) \neq 0 \tag{8.28}$$

Since:

$f_1(x_k) = 0.8x^2 + 0.2 \neq 0$ , that is no real root of $x$ can be found for $f_1(x) = 0$, we get $f_2(y) = 1$; series of values for $x_k$ and $y_k$ that will converge quadratically to the equilibrium point $x$ and $y$ . As a starter, pick the initial guess to be $x_0 = 0$, $y_0 = 0$ , we find $(x, y)$ to be:

$$(x, y) = \left( \frac{(0.467 + 0.8A\sin 0.2t)}{0.2}, \ -1.5 - \frac{0.467}{0.8} \right), \tag{8.29}$$

which yields a parametric equation for generating the equilibrium point for the Fitzhugh model.

From $V(t, x, y) = ax^2 + by^2$ we find that:

$$\dot{V}(t, x, y) = 2axy - 3ax + 2\,ax^2 - \frac{2a}{3} x^4$$

$$- 2ax\ A\ \sin 0.2t - \frac{2b}{9}xy + \frac{0.467(2)b}{9}y - \frac{1.6}{9}by^2 + \frac{1}{3}(0.8)by \qquad (8.30)$$

We also find that:

$$a_*\left(|x| + |y|\right)^2 \leq V(t, x, y) \leq a^*\left(|x| + |y|\right)^2$$

Where $a_* = \min\left[|a|, |b|\right]$, $a^* = \max\left[|a|, |b|\right]$ and also : $|V(t, x_1, y_1) - V(t, x_2, y_2)|$

$$\leq k_1|x_1 - x_2| + k_2|x_2 - y_2|$$

Where $k_1 = |a|\max(|x_1|, |x_2|)$, $k_2 = |b|\max(|y_1|, |y_2|)$ .We can show that :

$$V(t, x, y) \leq -c|x|V(t, x, y)$$

$$-\left(|x| + |y|\right)c\left(|x| + |y|\right)$$

Where $$c\left(|x| + |y|\right) = \left(x_2^2 + y_1^2\right)^{1/2}$$

This shows that Fitzhugh contains an equilibrium point that entertains uniform stability by Theorem 7.2.

## Avian Influenza Infection Models

Avian influenza (AI) is by far the most dangerous disease linking animal and human wellbeing today. Ten years ago, AI was a disease of poultry and wild birds of limited significance. Today, the emergence of a strain that can infect human beings through bird-to-human transmission and kill nearly 60% of those infected has changed this perspective Maia [12].

A mathematical model of Avian influenza which involves human Influenza is introduced to better understand the complex epidemiology of avian Influenza and the emergence of a pandemic strain. At present H5N1 avian influenza is a zoonotic disease where the transmission to human beings is through infected domestic birds. Since 2003, more than 500 people have been infected and nearly 60% of them have died. If the H5N1 virus becomes efficiently human-to-human transmittable, a pandemic will occur with potentially high mortality.

## Numerical Simulation of Bird-flu Epidemics

### *Model Formulation*

The proposed model describes the dynamics in the birds (chickens) and human being population subject to birds-flu. However, it is obvious that in human beings, the infection by this disease causes no permanent immunity and there is no effective vaccination [12] (Fig. **8.3**).

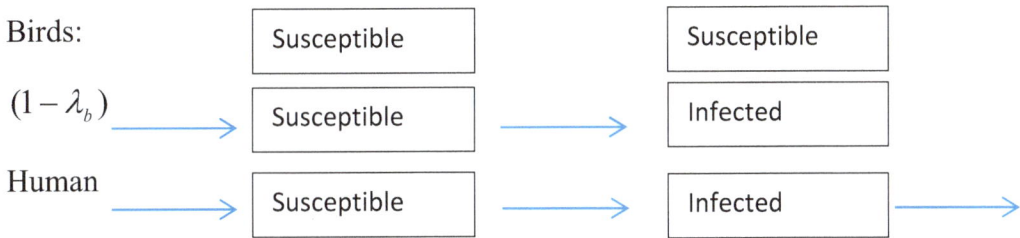

Birds:

$(1 - \lambda_b)$

Human

| Susceptible | Susceptible |
| Susceptible | Infected |
| Susceptible | Infected |

**Fig. (8.3).** Schematic description of the Model.

$$\frac{dS_b}{dt} = \beta N_B B - (A_B) M_B - \alpha B S_B - \frac{I_B}{N_B} \delta_B S_B$$

$$\frac{dI_B}{dt} = \alpha_B S_B \frac{I_B}{N_B} - (\delta_B + d_B) I_B + \lambda_B M_B$$

$$\frac{dS_H}{dt} = N_H \beta_B - \frac{\alpha_H S_H I_B}{NB} - \delta_H + I_H$$

$$\frac{dI_H}{dt} = \frac{\alpha_H S_H I_B}{N_B} - (\delta_H + d_H) I_H$$

$$(8.31)$$

Also, we assume that each of the total sub-populations $N_B$, $N_H > 0$ at t =0

Where:

$S_B$ = population of susceptible birds

$S_H$ = population of susceptible humans

$I_B$ = population of infectious birds

$I_H$ = population of infectious humans

$N_B$ = Total numbers of birds in the location of interest

$N_H$ =Total numbers of human beings in the location of interest

$\beta_B$ = Average birth rate in birds

$\beta_H$ = Average birth rate in human beings

$\lambda_B$ =Probability of infection in migrated birds

$M_B$ = Total number of migrated birds.

$\delta_H$ = Death rate in human beings

$\delta_B$ = Death rate in birds

$\alpha H$ =Infection transmission rate from birds to human beings

$\alpha_B$ = Infection transmission rate from birds to birds.

$d_B$ =Flu-induced death rate for birds

$d_N$ = Flu-induced death rate for human beings

$l$ = recovery rate for human beings

Analyze the behavior of the system of differential equations (8.30) by finding the equilibrium solutions and examine their stability.

The disease-free equilibrium model is obtained by setting the right-hand side of equation (8.30) to zero and letting all the infected terms to zero. This gives:

$$\frac{d\delta_B}{dt} = 0, \frac{dI_B}{dt} = 0, \frac{dS_H}{dt} = 0, \frac{dI_H}{dt} = 0$$

$$or\ \mathrm{E}_0 ; \left( \mathrm{S_B}^*, \mathrm{I_B}^*, \mathrm{S_H}^* \mathrm{I_H}^* \right)$$

This is:

$$N_B \beta_B + (1 - \lambda_B) M_B - \frac{\alpha_B S_B I_B}{N_B} - \delta_B S_B = 0$$

$$\frac{\alpha_B S_B I_B}{N_B} - (\delta_B + d_B) + \lambda_B M_B = 0$$

$$N_H \beta_H - \frac{\delta_B + d_B}{N_B} - \delta_H S_H + I_H = 0$$

$$\frac{\alpha_B S_B I_B}{N_B} - (\delta_B + d_H) I_H = 0$$

The linear stability of $\mathrm{E}_0$ is established using the Jacobian. This is done by obtaining the Jacobian matrix to the model equation (11).

The Jacobian of $S_B^*, I_B^*, S_H^*, I_H^*$ with respect to $\mathrm{S_B}, \mathrm{I_B}, \mathrm{S_H}, \mathrm{I_H}$ is :

$$\begin{vmatrix} \dfrac{\partial S_B^*}{\partial S_B} & \dfrac{\partial S_B^*}{\partial I_B} & \dfrac{\partial S_B^*}{\partial S_H} & \dfrac{\partial S_B^*}{\partial I_H} \\ \dfrac{\partial I_B^*}{\partial S_B} & \dfrac{\partial I_B^*}{\partial I_B} & \dfrac{\partial I_B^*}{\partial S_H} & \dfrac{\partial I_B^*}{\partial I_H} \\ \dfrac{\partial S_H^*}{\partial S_B} & \dfrac{\partial S_H^*}{\partial I_B} & \dfrac{\partial S_H^*}{\partial S_H} & \dfrac{\partial S_H^*}{\partial I_H} \\ \dfrac{\partial I_H^*}{\partial S_B} & \dfrac{\partial I_H^*}{\partial I_B} & \dfrac{\partial I_H^*}{\partial S_H} & \dfrac{\partial I_H^*}{\partial I_H} \end{vmatrix}$$

where

$S_B^*$ = the disease free equilibrium of susciptible birds

$\mathrm{I_B}^*$ = the disease free equilibrium of infection in birds

$\mathrm{S_H}^*$ = the disease free equilibrium of susceptible humands

$\mathrm{I_H}^*$ = the disease free equilibrium of infections in humans

That is:

$$S_B *(t) = N_B \beta_B + (1 - \lambda_B) M_B - \frac{\alpha_B S_B I_B}{N_B} - \delta_B S_B$$

$$I_B *(t) = \frac{\alpha_B S_B I_B}{N_B} - (\delta_B + d_B) + \lambda_B M_B$$

$$S_H *(t) = N_H \beta_H - \frac{\alpha_B S_B I_B}{N_B} - \delta_H S_H + I_H$$

$$I_H *(t) = \frac{\alpha_B S_B I_B}{N_B} - (\delta_B + d_H) I_H$$

Substituting (iv) into (iii), we have

$$\begin{vmatrix} -\left(\dfrac{\alpha_B I_B}{N_B} + \delta_B\right) & \dfrac{-\alpha_B S_B}{N_B} & 0 & 0 \\[3mm] \dfrac{\alpha_B I_B}{N_B} & \dfrac{\alpha_B S_B}{N_B} - (\delta_B + d_B) & 0 & 0 \\[3mm] 0 & \dfrac{-\alpha_H S_H}{N_B} & -\left(\dfrac{\alpha_H S_B}{N_B} + \partial_H\right) & \\[3mm] 0 & \dfrac{\alpha_H S_H}{N_B} & \dfrac{\alpha_H I_B}{N_B} - (\delta_H + d_H) & \end{vmatrix} \qquad \textbf{(8.32)}$$

In the above matrix, column (i) and column (4) are preserved while row operation is applied to other columns.

We obtain

$$\begin{pmatrix} -\delta_B & \dfrac{-\alpha_B S_B}{N_B} & 0 & 0 \\[3mm] 0 & \dfrac{\alpha_B S_B}{N_B} - (\delta_B + d_B) & 0 & 0 \\[3mm] 0 & \dfrac{-\alpha_H S_H}{N_B} & -\delta_H & 0 \\[3mm] 0 & \dfrac{\alpha_H S_H}{N_B} & 0 & -(\delta_H + d_H) \end{pmatrix}$$

$$S_B(t) = \frac{N_B\beta_H + (1-\lambda_B)M_B}{\delta B}$$

$$\delta_H(t) = \frac{N_H B_H}{\delta_H}$$

Putting $S_B(t)$ and $S_H(+)$ in (v)

We have

$$
\begin{pmatrix}
-\delta_B - \alpha B \dfrac{N_H\beta_H + (1-\lambda_B)M_B}{\delta_B} & 0 & 0 \\
0 & \alpha B\left(\dfrac{N_B\beta_H + (1-\lambda_B)M_B}{\delta_B}\right) - \delta_B + d_B & 0 & 0 \\
0 & -\dfrac{\alpha_H N_H\beta_H}{N_B\delta_H} & -\delta_H & 0 \\
0 & \dfrac{\alpha_H N_H\beta_H}{N_B\delta_H} & 0 & -(\delta_H + d_H)
\end{pmatrix}
$$

Apply row operation on $i^{th}$, so as to obtain characteristic roots of the above matrix.

We obtained:

$$
\begin{matrix}
-\delta_B - \lambda & -\alpha B\dfrac{NB\beta H + (1-\lambda B)M_B}{\delta_B} & 0 & 0 \\
0 & -(\delta_B + d_B) - \lambda & 0 & 0 \\
0 & 0 & -\delta_H - \lambda & -(\delta_H + d_H) \\
0 & \dfrac{\alpha_H N_H\beta_H}{N_B\delta_H} & 0 & -(\delta_H + d_H)
\end{matrix}
$$

$$-\delta_B - \lambda[((-\delta_B + d_B) - \lambda)(-(\delta_H + \lambda)(-\delta_H - d_H) - \lambda)] = 0$$

$$\lambda + \delta_B(\lambda + (\delta_B + d_B)(\lambda + \delta_H(\lambda + \delta_H + d_H))) = 0$$

$$\lambda_i = -\delta_B, \qquad \lambda_2 = _-(\delta_B + d_B)$$

$$\lambda_3 = -\delta_H \qquad \lambda_4 = -(\delta_4 + d_H)$$

When some values are substituted in the above eigenvalues and the results are negative, that means the Bird flu disease can be eradicated from bird-human beings' population.

Fig. (**8.4**) is the number of human H5N1 cases per year in 2004 to 2010 as predicted by the World Health Organization (WHO).

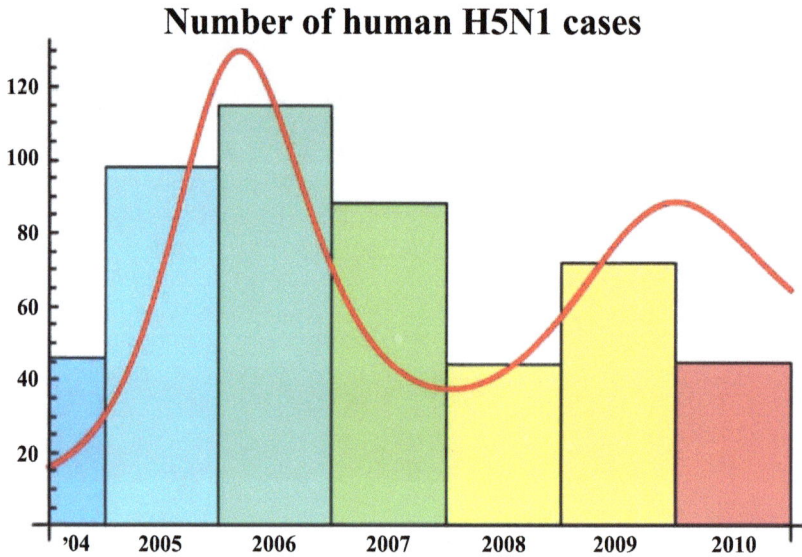

**Fig. (8.4).** Number of human cases H5N1 per year as predicted by WHO.

## Mathematical Model for Exploitation of Biological Predators

Kar and Chaudhri [19] studied the problem of combined harvesting of two competing species. Multi-species harvesting model and studies by Mesterten – Gibbons [23]. Oyelami and Ale studied fish-hyacinth model [17, 18].

We consider Kar-Chaudhuri model.

$$\frac{dx}{dt} = rx\left(1 - \frac{x}{K}\right) - \frac{m\,x\,y}{a + x}$$

$$\frac{dy}{dt} = sy\left(1 - \frac{y}{L}\right) + \frac{m\,x\,y}{a + x} \tag{8.33}$$

Where:

$x = x(t)$ = size of the prey population at time $t$.

$y = y(t)$ = size of the predictor population at time $t$

K = environmental carrying capacity.

L= environmental carrying capacity of the predator.

M= maximal relative increase of predator

$a$ = Michael's –Menten constant

$\alpha$ =Conversion factor (we assume $\alpha < 1$) since the whole biomass of the prey is not transformable to the predator.

R=   intrinsic growth rate of the prey

$S$ =intrinsic growth rate of the predator

If $f((x, y) = \dfrac{mxy}{a + x}$ it is Holling function, which is tropic or function response of the predator to the density of prey. Analytic solution to Kar-Chaudhari's model is intractable generally.  As test let us use the maple.

The problem can however be solved using numerical solvers in the Maple using the Runge-Kutta family and other solvers designed to handle stiff systems.

To determine the equilibrium state (biological equilibrium)

Set $$\frac{dx}{dt} = 0, \frac{dy}{dt} = 0$$

This involves solving

$$rx(1 - \frac{x}{K}) - \frac{mxy}{a + x} = 0$$
$$sy(1 - \frac{y}{L}) + \frac{mxy}{a + x} = 0$$

Using Matlab for particular values of r, s,m, a, K and L, the equilibrium state x* =x,y* =y in the above equations can be obtained using Matlab code

$$\gg solve('r * x * (1-\frac{x}{K})-(m * x * y)/(a+x)=0','s * y * (1-\frac{y}{L})-m * x * y/(a+x)=0');$$

Since, the analytic solution to model including the biological equilibrium cannot be obtained in closed form matlab.

Using Maple, we obtain the solution which depends on zero of complicated polynomial as follows:

> with(DETools) :

> with(plots) :

>
$$eq1 := r{\cdot}x{\cdot}\left(1-\frac{x}{l}\right)-\frac{m{\cdot}x{\cdot}y}{a+x}=0;$$

$$eq1 := rx\left(1-\frac{x}{l}\right)-\frac{mxy}{a+x}=0$$

>
$$eq2 := s{\cdot}y{\cdot}\left(1-\frac{y}{l}\right)+\frac{m{\cdot}x{\cdot}y}{a+x}=0;$$

$$eq2 := sy\left(1-\frac{y}{l}\right)+\frac{mxy}{a+x}=0$$

> solve({eq1, eq2}, [x, y]);

$$\left[[x=0,y=0],\,[x=0,y=l],\,[x=l,y=0],\,\Big[x=RootOf(rs\_Z^3+(2asr-lrs)\_Z^2+(a^2rs\right.$$

$$-2sarl+l^2m^2+l^2ms)\_Z-a^2lrs+al^2sm),y=$$

$$-\frac{1}{lm}\big(r\big(RootOf(rs\_Z^3+(2asr-lrs)\_Z^2+(a^2rs-2sarl+l^2m^2+l^2ms)\_Z$$

$$-a^2lrs+al^2sm)^2+RootOf(rs\_Z^3+(2asr-lrs)\_Z^2+(a^2rs-2sarl+l^2m^2$$

$$+l^2ms)\_Z-a^2lrs+al^2sm)\,a-l\,RootOf(rs\_Z^3+(2asr-lrs)\_Z^2+(a^2rs$$

$$-2sarl+l^2m^2+l^2ms)\_Z-a^2lrs+al^2sm)-la)\big)\Big]\Big]$$

We use qualitative techniques to study the behavior of the system rather than the solution to the model.

To investigate the stability of the model, we use the Lyapunov functional approach.

Now define
$$q_1 Ex = \frac{mxy}{a+x}, q_2 Ey = \frac{mxy}{a+x}$$

$q_i, i = 1, 2$ are the catchability coefficients of the two species [19].

Let

$$V(x,y) = \begin{vmatrix} r - \dfrac{2r}{K}x - \dfrac{amy}{(a+x)^2} - q_1 E & \dfrac{-mx}{(a+x)} \\[3mm] \dfrac{am\alpha y}{(a+x)^2} & s - \dfrac{2s}{L}y + \dfrac{amx}{(a+x)} - q_2 E \end{vmatrix} \qquad (8.34)$$

The biological equilibrium points are $(0,0),(x,0),(0,\bar{y}),(x^*,y^*)$ and:

$$V(x^*,y^*) = \begin{vmatrix} -\dfrac{r}{K}x^* + \dfrac{mx^* y^*}{(a+x)^2} & \dfrac{-mx^*}{(a+x^*)} \\[3mm] \dfrac{amy^*}{(a+x^*)^2} & \dfrac{-sy^*}{L} \end{vmatrix} \qquad (8.35)$$

It is proved see [19] that a necessary and sufficient condition for stable node is:

$$E > (\frac{r}{q_1}, \frac{s}{q_2})$$

The economic equilibrium state of the model is said to be achieved when $\dot{x} = 0, \dot{y} = 0$ and when TR (the total revenue obtained by selling the harvested biomass) equals TC (Total Cost for the effort devoted to harvesting). Then the revenue generated from selling the fish is:

$$R = p_1 q_1 xE + p_2 q_2 yE - CE$$

where $P_i$ are constant selling prices of x and y fishes $q\_\{i\}$ being the constant fishing effort costs.

For biological equilibrium set $\dot{x} = 0, \dot{y} = 0$

This occurs at the conic section:

$$\frac{r}{Kq_1}x^2 - \frac{s}{Lq_2}xy - \left(\frac{r}{q_1}\frac{s}{q_2} + \frac{ma}{q_2}\right)x + \left(\frac{m}{q_1} - \frac{as}{Lq_2}\right)y - \left(\frac{ar}{q_1} - \frac{as}{q_2}\right) = 0 \quad (8.36)$$

For economic equilibrium:

R = TR - TC = $(P_1q_1x + P_2q_2y - C)$ E $= 0$

which implies that $P_1q_1x+P_2q_2y-C=0$.               **(8.37)**

Equation (8.36) & (8.37) gives:

$$A_1x^2 + B_1x + C = 0,$$

Where:

$$A_1 = \frac{r}{Kq_1} + \frac{s}{Lq_2}\frac{P_1 q_1}{P_2 q_2} > 0$$

$$B_1 = \frac{s}{q_2} - \frac{r}{q_1} + \frac{ar}{Kq_1} - \frac{ma}{q_2} - \frac{Cs}{Lq_2}\frac{1}{P_2 q_2} - \left(\frac{n}{q_1} - \frac{as}{Lq_2}\right)\frac{P_1 q_1}{P_2 q_2}$$

$$C_1 = \frac{C}{P_2 q_2}\left(\frac{m}{q_1} - \frac{as}{Lq_2}\right) + \left(\frac{as}{q_2} - \frac{ar}{q_1}\right)$$

We can show that bionomic equilibrium exists if $B_1^2 - 4\,A_1 C_1 \geq 0$ and no bionomic equilibrium exists if:

$$B_1^2 - A_1 C_1 = 0.$$

## Population of Single Specie Model

Let the population density of a species in an environment be p and let $F(p)$ be a continuous function of $p$ on $\Omega = \{p : p > 0\}$. The general population model can be expressed as:

$\overset{\bullet}{p}(t) = pF(p)$, if $p^*$ is the equilibrium population; *i.e.* $p^*$ such that $p^* F(p^*) = 0$ [5] & [13].

## Definition 8.1

We say that a positive equilibrium $p^*$ which satisfies $p^* F(p^*) = 0$ is globally asymptotically stable if every solution of (1) which begins in the set $\Omega$ remains in it for all $t \geq 0$ and converges to $p^*$ as $t \to \infty$.

We use the direct method of Lyapunov to establish results on the global asymptotic stability (GAS) of equilibria. To establish GAS for a single specie population, we need a Lyapunov function $V(p^*)$ with the following properties:

(i) $V(p)$ must be positive definite such that $V(p^*) = 0$;

(ii) $V(p) \to \infty$ as $p \to 0^+$ and as $p \to \infty$.

(iii) $\overset{\bullet}{V}(p) = \left(\dfrac{\partial V}{\partial p}\right) pF(p)$ is negative definite for all $p > 0$

The positive equilibrium $p^*$ of equation (8.1) is globally asymptotically stable if there exists a Lyapunov function $V(p)$ such that $\overset{\bullet}{V}(p)$ does not vanish identically along a nontrivial solution $p(t) \neq p^*$, an excellent choice of Lyapunov function See Goh [5] is $V(p) = \displaystyle\int_{p^*}^{p} \dfrac{h(s)}{g(s)} ds$,

where $h(s)$ and $g(s)$ are continuous functions such that: $h(s) < 0$ for all $0 \leq s < p^*, h(p^*) > 0$ for all $0 < s < \infty$ and $g(s) > 0$ for all $s > 0$.

Furthermore, $g(s)$ and h(s) are chosen so that $V(N) \to \infty$ as $N \to 0^+$ and as $N \to \infty$.

## Example 8.1 Goh [5]

Let $h(s) = s - p^*$ and $g(s) = s$.

We get $V(p) = p - p^* - p^* \ln(p/p^*)$ and $\dot{V}(p) = (p - p^*)F(p)$, it follows that the positive equilibrium $p^*$ of (8.1) is globally asymptotically stable if $F(p) > 0$ for all $p^* < p < \infty$.

The idea of (GAS) can be extended to multispecies populations as follows:

Let $p_i$ be the population of the $i^{\text{th}}$ specie in a community of $n$ interacting species. Suppose the community population

$$\dot{p}_i = p_i F(p_1, p_2, p_{3,...,}p_n), i = 1, 2, 3, ..., m,$$

where $F_1, F_2, F_3, ..., F_m$ are continuous functions. In the positive octant $\Omega = \{p_i : p_i > 0, i = 1, 2, 3, ..., m\}$, let $p^*$ be an equilibrium population which satisfied $F_i(p) = 0$ for $i = 1, 2, ..., m$. We say that $p^*$ is a positive equilibrium if $p^* > 0, i = 1, 2, ..., m$. We can extend the idea of Lyapunov stability for single population to multispecies populations as follows:

Construct a Lyapunov function $V(p)$, which is a continuous function of $p$ such that:

i.   $V(p)$ is positive definite such that $V(p^*) = 0$
ii.  $V(p) \to \infty$ as $p_i \to 0^+$ and $p_i \to \infty$ for $i = 1, 2, .., m$
iii. The time derivative of $V(p)$ along every solution of (MSP) is negative definite.

Then, we have
$$\dot{V}(p) = \sum_{i=1}^{m} \left( \frac{\partial V}{\partial p_i} \right) p_i F(p_i).$$

We can construct a Lyapunov function for MSP by choosing $V(p) = \sum_{i=1}^{m} c_i V_i(p)$.

Here $c_i, i = 1, 2, ..., m$ are positive constants and must be chosen such that $\dot{V}(p)$ is negative definite. A simple example is $V(p) = \sum_{i=1}^{m} \int_{p_i^*}^{p_i} \left[ \frac{h_i(s)}{g_i(s)} \right] ds$, where $h_i(s)$ and $g_i(s)$ are functions of single specie populations. We can for example choose $g_i(s) = s$ and $h_i(s) = s - p_i^*$.

Therefore $\dot{V}(p) = \sum_{k=1}^{m} c_i \left( p_i - p_i^* \right) F_i(p)$, which is negative semi definite in the population in $\Omega$ and $\dot{V}(p)$ does not vanish identically along a nontrivial solution of MSP except at $p = p^*$

**Example** 8.2 Goh [5]

Consider the linear food chain model:

$$\dot{p}_1 = p_1(b_1 - a_{12}p_2)$$
$$\dot{p}_2 = p_2(-d_1 + e_1 a_{12} p_1 - a_{23} p_3)$$
$$\dot{p}_3 = p_3(-d_3 + e_3 a_{23} p_2 - a_{33} p_3)$$

where $b_1, d_1, d_3, e_1, a_{12}, a_{23}$ and $a_{33}$ are positive constants.

The equilibrium points are:

$$p_2^* = \frac{b_1}{a_{12}}, p_3^* = \frac{(e_3 a_{23} p_2^* - d_3)}{a_{33}} \text{ and } p_1^* = \frac{a_{23} p_3^* + d_1}{e_1 a_{12}}$$

Suppose $p^*$ is positive if $c_2 = 1/e_2$ and $c_3 = 1/(e_2 e_3)$. Then $\dot{V}(p)$ reduces to

$$\dot{V}(p) = -\left( \frac{a_{33}}{e_2 e_3} \right) \left( p_3 - p_3^* \right)^2$$

Therefore, the equilibrium $p^*$ is global asymptotically stable.

**Maple Examples**

The Linearize command computes the linearization of equations about an operating point $(x_0, u_0)$ specified by the limit point. It is assumed that the equations can be reduced to the form:

$$\frac{d}{dt} x(t) = f(x(t), u(t))$$

$$y(t) = g(x(t), u(t))$$

Where:

$x(t)$  Represents the state variables of the equations converted to the first-order form;

$u(t)$  Represents the input variables specified by b or c;

$y(t)$  Represents the output variables specified by y;

$f$ and $g$  are nonlinear functions to be linearized;

$t$ is a continuous time variable as specified by the continuoustimevar option in DynamicSystem[SystemOptions].

The Linearize command simplifies equations to the above form and computes the following linear model.

$$\frac{d}{dt} x(t) = a(x(t) - x_0) + b(u(t) - u_0) + f(x_0, u_0)$$

$$y(t) = c(x(t) - x_0) + d(u(t) - u_0) + g(x_0, u_0) ,$$

where  $a = \frac{\partial}{\partial x} f$,  $b = \frac{\partial}{\partial u} f$,  $c = \frac{\partial}{\partial x} g$,  and  $d = \frac{\partial}{\partial u} g$   are the Jacobian matrices evaluated at linpoint.

$with(DynamicSystems)$ :

> $sys1 := \left[ \frac{d}{dt} x_1(t) = x_2(t)^2 - 4, \frac{d}{dt} x_2(t) = x_1(t) - 1 + \sin(2 \cdot t) + b(t), y(t) = x_1(t) + x_2(t) \right]$

$sys1 := \left[ \frac{d}{dt} x_1(t) = x_2(t)^2 - 4, \frac{d}{dt} x_2(t) = x_1(t) - 1 + \sin(2 t) + b(t), y(t) = x_1(t) + x_2(t) \right]$

> $EquilibriumPoint\left(sys1, [b(t)], constraints = [0 < x_1(t)], initialpoint = [b(t) = 0, x_1(t) = 2,\right.$
>    $\left.x_2(t) = 4]\right);$

$$[x_1(t) = 1.29366189940414, x_2(t) = -1.99999999997961], \left[\frac{d}{dt} x_1(t) = \right.$$

$$\left. -8.15578715673837\ 10^{-11}, \frac{d}{dt} x_2(t) = 2.38878716807278\ 10^{-9}\right], [b(t) =$$

$$-0.706338100595861, t = 0.212695081293687], [y(t) = -0.706338100575471]$$

> $sys2 := \left[\frac{d}{dt} x_1(t) = 2 \cdot x_2(t)^2 - 8, \frac{d}{dt} x_2(t) = 2 \cdot x_1(t) - 2 + c(t), y(t) = 2 \cdot x_1(t) + 3 \cdot x_2(t)\right];$

$$sys2 := \left[\frac{d}{dt} x_1(t) = 2 x_2(t)^2 - 8, \frac{d}{dt} x_2(t) = 2 x_1(t) - 2 + c(t), y(t) = 2 x_1(t) + 3 x_2(t)\right]$$

> $eq\_point1 := EquilibriumPoint\left(sys2, [c(t)], constraints = [0 < x_1(t)], initialpoint = [c(t) = 0,\right.$
>    $\left. x_1(t) = 1, x_2(t) = 3]\right);$

$$eq\_point1 := [x_1(t) = 1., x_2(t) = -2.00000000000026], \left[\frac{d}{dt} x_1(t) = 2.07478478841949\ 10^{-12},\right.$$

$$\left.\frac{d}{dt} x_2(t) = 9.968367279\ 10^{-318}\right], [c(t) = 9.968367279\ 10^{-318}], [y(t) =$$

$$-4.00000000000078]$$

> $lin\_point1 := \left[op\left(eq\_point1_1\right), op\left(eq\_point1_3\right)\right];$

$$lin\_point1 := \left[x_1(t) = 1., x_2(t) = -2.00000000000026, c(t) = 9.968367279\ 10^{-318}\right]$$

> $lin\_model1a := Linearize(sys2, [c(t)], [y(t)], lin\_point1);$

$$lin\_model1a := \begin{array}{|l} \qquad\qquad \textbf{State Space} \\ \qquad\qquad \text{continuous} \\ \text{1 output(s); 1 input(s); 2 state(s)} \\ \qquad \text{inputvariable} = [u_1(t)] \qquad, [x_1(t) = x_1(t) - 1., x_2(t) = x_2(t) \\ \qquad \text{outputvariable} = [y_1(t)] \\ \qquad \text{statevariable} = [x_1(t), x_2(t)] \end{array}$$

$$+ 2.000000000], \left[u_1(t) = c(t) - 9.968367279\ 10^{-318}\right], [y_1(t) = y(t) + 4.]$$

> *PrintSystem*(*lin_model1a*₁)

$$
\left|
\begin{array}{l}
\textbf{State Space} \\
\text{continuous} \\
\text{1 output(s); 1 input(s); 2 state(s)} \\
\text{inputvariable} = \left[\, u_1(t)\,\right] \\
\text{outputvariable} = \left[\, y_1(t)\,\right] \\
\text{statevariable} = \left[\, x_1(t), x_2(t)\,\right] \\
a = \begin{bmatrix} 0 & -8.000000000 \\ 2. & 0 \end{bmatrix} \\
b = \begin{bmatrix} 0 \\ 1 \end{bmatrix} \\
c = \begin{bmatrix} 2. & 3. \end{bmatrix} \\
d = \begin{bmatrix} 0 \end{bmatrix}
\end{array}
\right.
$$

> *lin_model1b* := *Linearize*(*sys2*, [*c*(*t*)], [*y*(*t*)], *lin_point1*,'*equilibriumtolerance*'= 1. 10⁻¹⁰)

$$
lin\_model1b := \left[
\begin{array}{c}
\textbf{State Space} \\
\text{continuous} \\
\text{1 output(s); 1 input(s); 2 state(s)} \\
\text{inputvariable} = \left[\, u_1(t)\,\right] \qquad , \left[ x_1(t) = x_1(t) - 1., x_2(t) = x_2(t) \right.\\
\text{outputvariable} = \left[\, y_1(t)\,\right] \\
\text{statevariable} = \left[\, x_1(t), x_2(t)\,\right]
\end{array}
\right.
$$

$$
\left. + 2.000000000\right], \left[ u_1(t) = c(t) - 9.968367279\ 10^{-318}\right], \left[ y_1(t) = y(t) + 4.\right]
$$

> *PrintSystem*(*lin_model1b*₁)

**State Space**

continuous

1 output(s); 1 input(s); 2 state(s)

inputvariable $= \left[ u_1(t) \right]$

outputvariable $= \left[ y_1(t) \right]$

statevariable $= \left[ x_1(t), x_2(t) \right]$

$$a = \begin{bmatrix} 0 & -8.000000000 \\ 2. & 0 \end{bmatrix}$$

$$b = \begin{bmatrix} 0 \\ 1 \end{bmatrix}$$

$$c = \begin{bmatrix} 2. & 3. \end{bmatrix}$$

$$d = \begin{bmatrix} 0 \end{bmatrix}$$

> $lin\_model1c := Linearize\left( sys2, \left[ c(t) \right], \left[ y(t) \right], lin\_point1, 'equilibriumtolerance' = 1. \, 10^{-10}, \right.$
> $\left. 'checkpoint' = false \right)$

$lin\_model1c :=$

$\qquad$ **State Space**

$\qquad$ continuous

$\qquad$ 1 output(s); 1 input(s); 2 state(s)

$\qquad$ inputvariable $= \left[ u_1(t) \right]$    , $\left[ x_1(t) = x_1(t) - 1., x_2(t) = x_2(t) \right.$

$\qquad$ outputvariable $= \left[ y_1(t) \right]$

$\qquad$ statevariable $= \left[ x_1(t), x_2(t) \right]$

$+ 2.000000000 \Big], \left[ u_1(t) = c(t) - 9.968367279 \, 10^{-318} \right], \left[ y_1(t) = y(t) + 4. \right]$

Example of an Inverted pendulum on a moving cart (Fig. **8.5**):

## Variables

$\theta(t)$          counter-clockwise angular displacement of the pendulum from the upright position

$\phi(t)$          angular velocity of the pendulum, $\phi(t) = \dfrac{d}{dt}\,\theta(t)$

$x(t)$          position of the cart

$v(t)$          velocity of the cart, $y(t) = \dfrac{d}{dt}\,x(t)$

$u(t)$          horizontal force applied to the cart

## Parameters

$L$          half-length of pendulum

$m$          mass of the pendulum

$M$          mass of the cart

$g$          gravitational constant $\left(9.8\ \dfrac{m}{s^2}\right)$

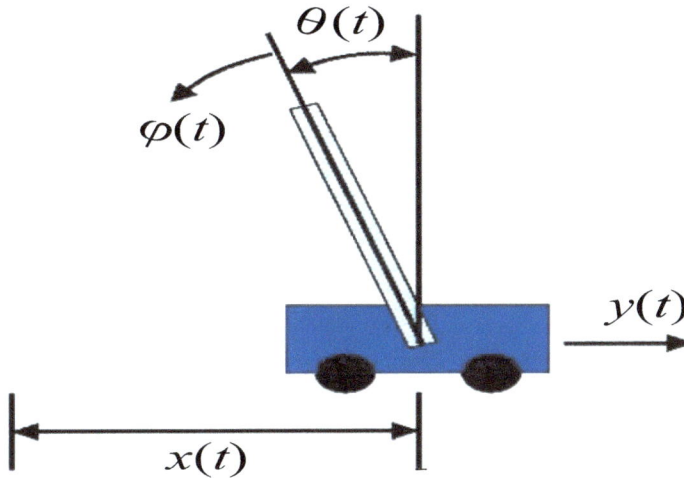

**Fig. (8.5).** Movement of a cart with application controls forces.

\>

$$sys3 := \left| \frac{d}{dt} x(t) = y(t), \frac{d}{dt} \theta(t) = \phi(t), \frac{d}{dt} y(t) = \right.$$

$$-\frac{-3\cos(\theta(t))\sin(\theta(t))g - 2u(t) + 2mL\sin(\theta(t))\phi(t)^2}{-3\cos(\theta(t))m + 2M + 2m}, \frac{d}{dt}\phi(t) =$$

$$\left. -\frac{3\left(-\sin(\theta(t))gM - \sin(\theta(t))gm - mu(t) + m^2L\sin(\theta(t))\phi(t)^2\right)}{(-3\cos(\theta(t))m + 2M + 2m)mL} \right|$$

$$sys3 := \left| \frac{d}{dt} x(t) = y(t), \frac{d}{dt} \theta(t) = \phi(t), \frac{d}{dt} y(t) = \right.$$

$$-\frac{-3\cos(\theta(t))\sin(\theta(t))g - 2u(t) + 2mL\sin(\theta(t))\phi(t)^2}{-3\cos(\theta(t))m + 2M + 2m}, \frac{d}{dt}\phi(t) =$$

$$\left. -\frac{3\left(-\sin(\theta(t))gM - \sin(\theta(t))gm - mu(t) + m^2L\sin(\theta(t))\phi(t)^2\right)}{(-3\cos(\theta(t))m + 2M + 2m)mL} \right|$$

\>
$$lin\_point3 := \left[\phi(t) = 0, x(t) = 0, y(t) = 0, \theta(t) = 0, u(t) = 0\right]$$

$$lin\_point3 := \left[\phi(t) = 0, x(t) = 0, y(t) = 0, \theta(t) = 0, u(t) = 0\right]$$

\>
$$lin\_model3 := Linearize(sys3, [u(t)], [\phi(t), x(t), y(t), \theta(t)], lin\_point3)$$

$$lin\_model3 := \left[ \begin{array}{c} \textbf{State Space} \\ \text{continuous} \\ \text{4 output(s); 1 input(s); 4 state(s)} \\ \text{inputvariable} = \left[ u_1(t) \right] \\ \text{outputvariable} = \left[ y_1(t), y_2(t), y_3(t), y_4(t) \right] \\ \text{statevariable} = \left[ x_1(t), x_2(t), x_3(t), x_4(t) \right] \end{array} \right. , \left[ x_1(t) = \phi(t), x_2(t) = \theta(t), x_3(t) \right.$$

$$= x(t), x_4(t) = y(t) \right], \left[ u_1(t) = u(t) \right], \left[ y_1(t) = \phi(t), y_2(t) = \theta(t), y_3(t) = x(t), y_4(t) = y(t) \right]$$

> *PrintSystem*$\left( lin\_model3_1 \right)$

$$\left[ \begin{array}{l} \textbf{State Space} \\ \text{continuous} \\ \text{4 output(s); 1 input(s); 4 state(s)} \\ \text{inputvariable} = \left[ u_1(t) \right] \\ \text{outputvariable} = \left[ y_1(t), y_2(t), y_3(t), y_4(t) \right] \\ \text{statevariable} = \left[ x_1(t), x_2(t), x_3(t), x_4(t) \right] \\[2mm] a = \begin{bmatrix} 0 & -\dfrac{3.\,(-gM - gm)}{(-m + 2.\,M)\,mL} & 0 & 0 \\ 1 & 0 & 0 & 0 \\ 0 & 0 & 0 & 1 \\ 0 & \dfrac{3.\,g}{-m + 2.\,M} & 0 & 0 \end{bmatrix} \\[6mm] b = \begin{bmatrix} \dfrac{3.}{(-m + 2.\,M)\,L} \\ 0 \\ 0 \\ \dfrac{2.}{-m + 2.\,M} \end{bmatrix} \\[6mm] c = \begin{bmatrix} 1 & 0 & 0 & 0 \\ 0 & 1 & 0 & 0 \\ 0 & 0 & 1 & 0 \\ 0 & 0 & 0 & 1 \end{bmatrix} \\[4mm] d = \begin{bmatrix} 0 \\ 0 \\ 0 \\ 0 \end{bmatrix} \end{array} \right]$$

## Prey -Predator Model

We consider a prey-predator population in a given ecological setup.

> $$eq1 := r \cdot x \cdot \left( 1 - \frac{x}{k} \right) - \frac{m \cdot x \cdot y}{a + x};$$

$$eq1 := rx\left(1 - \frac{x}{k}\right) - \frac{mxy}{a+x}$$

> 

$$eq2 := s{\cdot}y{\cdot}\left(1 - \frac{y}{L}\right) - \frac{m{\cdot}x{\cdot}y}{a+x};$$

$$eq2 := sy\left(1 - \frac{y}{L}\right) - \frac{mxy}{a+x}$$

> 

$$solve(\{eq1, eq2\}, [x, y]);$$

$$\left[[x=0, y=0], [x=0, y=L], [x=k, y=0], \left[x = RootOf\left(rs\_Z^3 + (2\,asr - krs)\_Z^2 + (\right.\right.\right.$$

$$-Lkm^2 + Lkms + a^2\,rs - 2\,akrs)\_Z + Laskm - a^2\,krs), y =$$

$$-\frac{1}{km}\left(r\left(RootOf\left(rs\_Z^3 + (2\,asr - krs)\_Z^2 + (-Lkm^2 + Lkms + a^2\,rs\right.\right.\right.$$

$$- 2\,akrs)\_Z + Laskm - a^2\,krs\right)^2 + RootOf\left(rs\_Z^3 + (2\,asr - krs)\_Z^2 + (\right.$$

$$-Lkm^2 + Lkms + a^2\,rs - 2\,akrs)\_Z + Laskm - a^2\,krs\right) a - kRootOf\left(rs\_Z^3\right.$$

$$+ (2\,asr - krs)\_Z^2 + (-Lkm^2 + Lkms + a^2\,rs - 2\,akrs)\_Z + Laskm - a^2\,krs\right)$$

$$\left.\left.\left.\left. - ka\right)\right)\right]\right]$$

> restart

>  # Enter the parameters

> r := 0.03;

r := 0.03

> s := 0.01;

s := 0.01

> k := 50000;

k := 50000

> L := 10000;

$L := 10000$

> $m := 0.005;$

$m := 0.005$

> $a := 1.5;$

$a := 1.5$

# Enter the equilibrium equations

>
$$eq1 := r \cdot x \cdot \left(1 - \frac{x}{k}\right) - \frac{m \cdot x \cdot y}{a + x};$$

$$eq1 := 0.03\, x \left(1 - \frac{x}{50000}\right) - \frac{0.005\, x y}{1.5 + x}$$

>
$$eq2 := s \cdot y \cdot \left(1 - \frac{y}{L}\right) - \frac{m \cdot x \cdot y}{a + x};$$

$$eq2 := 0.01\, y \left(1 - \frac{y}{10000}\right) - \frac{0.005\, x y}{1.5 + x}$$

# Find the equilibrium points

> $solve(\{eq1, eq2\}, [x, y]);$

$[[x = 0., y = 0.], [x = 0., y = 10000.], [x = 50000., y = 0.], [x = 847.7027317, y = 5008.831813],$
$[x = 49152.29458, y = 5000.152582], [x = -2.997309516, y = -8.984395643]]$

# Input the differential equations and initial conditions

>
$$ode1 := diff(x(t), t) = 0.05\, x(t) \cdot \left(1 - \frac{x(t)}{50000}\right) - \frac{0.005 \cdot x(t) \cdot y(t)}{1.5 + x(t)};$$

$$ode1 := \frac{d}{dt}\, x(t) = 0.05\, x(t) \left(1 - \frac{x(t)}{50000}\right) - \frac{0.005\, x(t)\, y(t)}{1.5 + x(t)}$$

>
$$ode2 := diff(y(t), t) = 0.06 \cdot y(t) \cdot \left(1 - \frac{y(t)}{40000}\right) - \frac{0.001\, x(t) \cdot y(t)}{2.5 + x(t)};$$

$$ode2 := \frac{d}{dt}\, y(t) = 0.06\, y(t) \left( 1 - \frac{y(t)}{40000} \right) - \frac{0.001\, x(t)\, y(t)}{2.5 + x(t)}$$

> $ic := x(0) = 5000, y(0) = 1000;$

$ic := x(0) = 5000, y(0) = 1000$

\# Find the numerical solution to the model

> $sol := dsolve(\{ode1, ode2, ic\}, [x(t), y(t)], numeric);$

$sol := \textbf{proc}(x\_rkf45) \;...\; \textbf{end proc}$

\# plot the solution curve

> $plots[odeplot](sol, [[t, x(t), color = black], [t, y(t), color = blue]], 0..100, labels = [t,$
  $'Number\ of\ species']);$

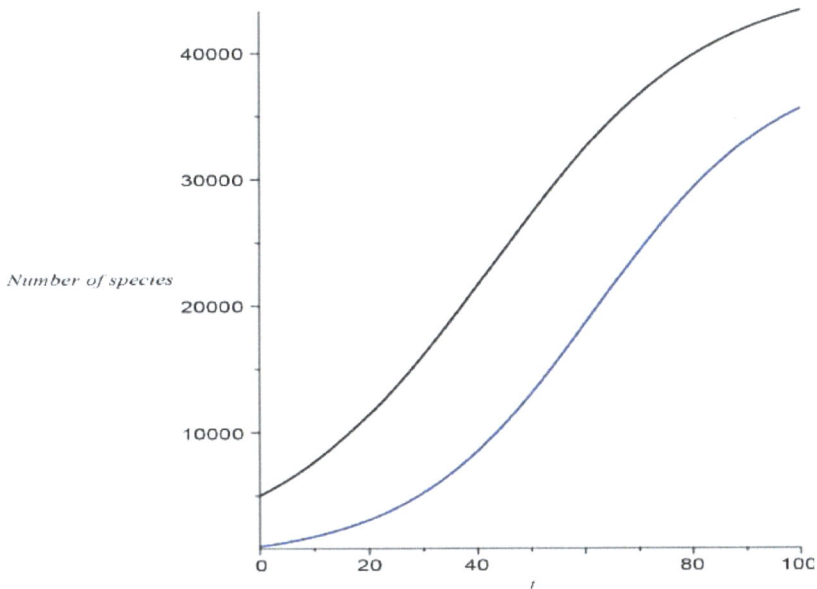

**Fig. (8.6).** Plot of the population of Preys and Predators.

Fig. (**8.6**) is Preys and Predators coexisting in a given ecological community. From the graph, the predators do not poise problem to the prey population.

## Numerical Simulation to the Bird flu model

## Notation:

$X(t)$= population of susceptible birds

$Y(t)$ = population of susceptible humans

$Z(t)$ = population of infectious birds

$W(t)$= population of infectious humans

> *with(ODETools)* :

> *with(plots)* :

> $\beta_1$ := 0.1; $n_1$ := 0.01; $m_1$ := 0.02 ; $\alpha_1$ := 0.01; $d_1$ := 0.012 # *enter the values of parameters;*

$\beta_1$ := 0.1

$n_1$ := 0.01

$m_1$ := 0.02

$\alpha_1$ := 0.01

$d_1$ := 0.012

## > # The bird flu equations

$de1 := diff(X(t), t) = beta[1] \cdot n[1] - n[1] \cdot m[1] - alpha[1] \cdot X(t) - \dfrac{1}{n[1]} \cdot alpha[1] \cdot X(t) \cdot Z(t);$

$$de1 := \frac{d}{dt} X(t) = 0.0008 - 0.01\, X(t) - 1.\, X(t)\, Z(t)$$

> $de2 := diff(Z(t), t) = \dfrac{alpha[1] \cdot Y(t) \cdot Z(t)}{n[1]} - (alpha[1] + d[1]) \cdot Z(t) + alpha[1] \cdot m[1];$

$$de2 := \frac{d}{dt} Z(t) = 1.000000000\, Y(t)\, Z(t) - 0.022\, Z(t) + 0.0002$$

> $$de3 := diff(Y(t), t) = beta[1] \cdot n[1] - alpha[1] \cdot Y(t) \cdot Z(t) - alpha[1] \cdot m[1];$$

$$de3 := \frac{d}{dt} Y(t) = 0.0008 - 0.01\, Y(t)\, Z(t)$$

> $$de4 := diff(W(t), t) = \frac{alpha[1] \cdot Y(t) \cdot Z(t)}{n[1]} - (alpha[1] + d[1]) + W(t);$$

$$de4 := \frac{d}{dt} W(t) = 1.000000000\, Y(t)\, Z(t) - 0.022 + W(t)$$

> $eq := \{de1, de2, de3, de4\};$

$$eq := \left\{ \frac{d}{dt} W(t) = 1.000000000\, Y(t)\, Z(t) - 0.022 + W(t), \frac{d}{dt} X(t) = 0.0008 - 0.01\, X(t) \right.$$
$$- 1.\, X(t)\, Z(t), \frac{d}{dt} Y(t) = 0.0008 - 0.01\, Y(t)\, Z(t), \frac{d}{dt} Z(t) = 1.000000000\, Y(t)\, Z(t)$$
$$\left. - 0.022\, Z(t) + 0.0002 \right\}$$

> $ic := \left\{ X(0) = 1 + \beta_1, Z(0) = 12 - n_1, Y(0) = 5 + n_1, W(0) = 2 + \alpha_1 + d_1 \right\};$

$$ic := \{ W(0) = 2.022, X(0) = 1.1, Y(0) = 5.01, Z(0) = 11.99 \}$$

> $dsol2 := dsolve(eq \textbf{ union } ic, numeric)$

$dsol2 := \textbf{proc}(x\_rkf45) \ ... \ \textbf{end proc}$

> $dsol2(0);$ # find the numerical solution at t=0;

$[t = 0., W(t) = 2.02200000000000, X(t) = 1.10000000000000, Y(t) = 5.01000000000000, Z(t)$
$= 11.9900000000000]$

> $dsol2(0.012)$

$[t = 0.012, W(t) = 2.79286097930696, X(t) = 0.948312095212564, Y(t)$
$= 5.00258671642338, Z(t) = 12.7290287940541]$

> $plots[\,odeplot\,](dsol2,[[t,W(t),color=blue],[t,Y(t),color=red],[t,Z(t),color=green\,],[t,$
> $X(t),color=black]],0\mathbin{..}1,labels=[t,""\,]);$

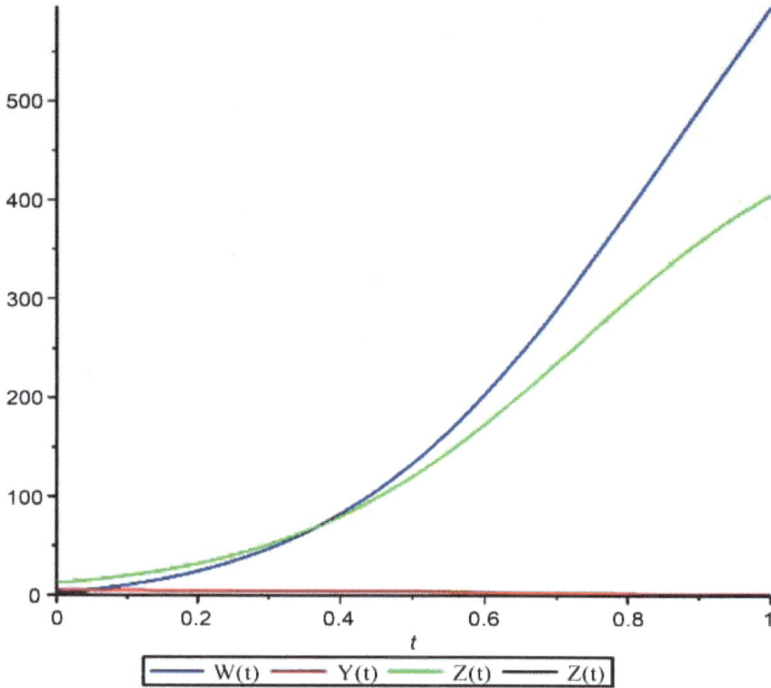

**Fig. (8.7).** The graph of solution to Bird flu model.

Fig. (**8.7**) is a graph of solution to Bird flu model, where $X(t), Y(t), Z(t)$ and $W(t)$ are different population groups described in the Notation section. In the Fig. (**8.3**), the population of susceptible humans and that of susceptible birds is persistent. That is, their populations continuously decrease and tending to zero.

In the following Example, we compute the proportion of the population in the environment with time.

$$omag = X(t) + Y(t) + Z(t) + W(t) = N$$

N is the total population of species in the environment. Therefore:

$\dfrac{X(t)}{omag}$ = proportion of susceptible birds

$\dfrac{Y(t)}{omag}$ = proportion of susceptible humans

$\dfrac{Z(t)}{omag}$ = proportion of infectious birds

$\dfrac{W(t)}{omag}$ = proportion of infectious humans

> $omagW := W(t) + X(t) + Y(t) + Z(t);$

color="NavyBlue"

thickness=3

filled=[color="Blue", transparency=0.5];

$$omag := W(t) + X(t) + Y(t) + Z(t)$$

>

$$plots[odeplot]\left( dsol2, \left[ \left[ t, \frac{W(t)}{omag}, color=red \right], \left[ t, \frac{Y(t)}{omag}, color=blue \right], \left[ t, \frac{Z(t)}{omag}, color \right. \right.$$
$$\left. \left. = yellow \right], \left[ t, \frac{X(t)}{omag}, color=cyan \right] \right], 0..1, labels = [t, ""] \right)$$

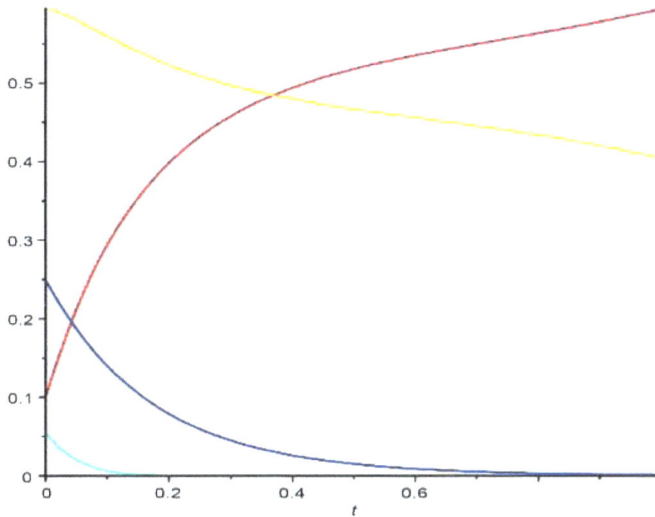

Fig. (8.8). The graph of proportion of species present in the environment in Bird flu model for $0 \leq t \leq 1$ week.

**Fig. (8.8)** shows proportion of infectious humans are continuously increasing with time (red color curve) while susceptible birds and susceptible humans (blue color curve) are persistent. Proportion of infectious birds (yellow color curve) is decreasing but not steadily like susceptible species. Fig (**8.9**) is the same simulation carried out for an extended period of five weeks instead of a week period in Fig. (**8.8**) Similar interpretation of the result can be made on the dynamic of disease in the environment for five weeks.

> $plots[\,odeplot]\left(dsol2,\left[\left[t,\dfrac{W(t)}{omag},color=red\right],\left[t,\dfrac{Y(t)}{omag},color=blue\right],\left[t,\dfrac{Z(t)}{omag},color\right.\right.$
> $\left.\left.=yellow\right],\left[t,\dfrac{X(t)}{omag},color=cyan\right]\right],0..5,labels=[t,""]\right)$

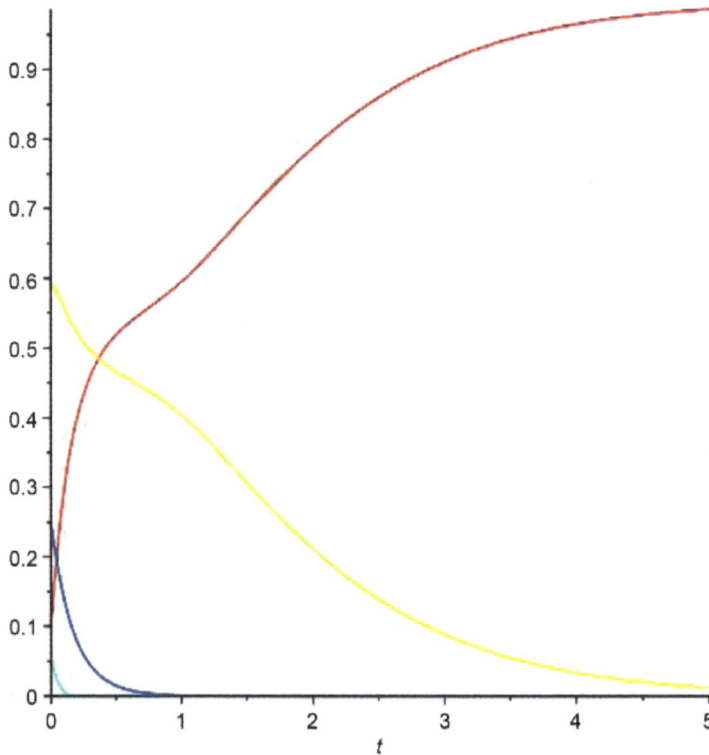

**Fig. (8.9).** The graph of proportion of species present in the environment in Bird flu model for $0 \le t \le 5$ weeks.

## CONCLUSION

Neural networks are primarily concerned with modeling the activity of the brain, its behavioral processes, and the application of these models to computers and related technologies. In the present age of big data, machine learning and artificial intelligence, modeling human activities *via* neural networks is the crux of Research these days. We considered a variety of popular neural network models from the family of differential equations and use them to characterize a number of state vectors which converge to stable equilibrium states using iterative maps. The lyapunov stable technique is used to investigate the stability of equilibrium points of Grossberg, Hopfield, Fitzhugh-Nagumo models. Also worthy to mention, is the study on epidemiology of avian flu and bird flu *via* models. The condition for existence of disease free equilibrium is obtained with associated stability properties. Also studied are combined harvesting problems for two competing species, multi-species harvesting problem, population dynamic of single species and multi-species problems. The stability properties of these models are investigated using lyapunov's stability technique with interesting results obtained. A simple method for constructing a Lyapunov function is presented.

## REFERENCES

[1]    Cichocki, and R. Unbehauen, *Neural Network for Optimization and Signal Processing..* Academic Press: London, New York, 1993.

[2]    J.M. Cushing, R.F. Constantino, D. Brian, R. A Desharnais, and M. A. Shandeke, "Nonlinear dynamics: Models, experiment and data", *J. Theor boil,* vol. 194, pp. 1-9, 1998.

[3]    G. Birkhoff, and Rota. Gian-Carlo, *Ordinary Differential Equations.* Wiley, 1989.

[4]    A. Tagliarini Gene, Christ J. Jury, and W.P. Edward, *Neural Networks: Theory and Modeling Optimization using Neural Networks.,* vol. 40. Institute of Electrical Engineering Applications on Computer , 1991no. 12, .

[5]    B.S. Goh, *Stability of some Multispecies Population Models in Modeling and Differential Equations in Biology .* Lecture Notes in Pure and Applied Mathematics Marcel Dekker, Inc. Publication: New York, 1980, pp. 209-216.

[6]    Stephen. Grossberg, *Studies of Mind and Brain: Neural Principles of Learning, Perception, Development, Cognition, and Motor Control..* Reidel Press, 1987.

[7]    P.V. Gupta, and P.C. Dhar, *Network Analysis and Synthesis.* Dhan Pat Raj Publication: Delhi, India, 2006.

[8]    J.J. Hopfield, "Neurons, dynamics and computation", *Phys. Today,* vol. 47, no. 2, pp. 40-46, 1994.
       http://dx.doi.org/10.1063/1.881412

[9]    N. Decluris, *Neural Network,* 7th Edition McGraw-ill Encyclopedia of Science and Technology, 1992, pp. 671-673.

[10]   H. Naylor Thomas, D. Joseph, L.B. Balinfty, and K. Kongehuk, *Computer.* John Wiley and Son Inc.: U.S.A, 1966.

[11]   K. Hale Jack, *Ordinary Differential Equations.* Wiley-Interscience: New York, London, Sydney, Toronto, 1969.

[12]   M. Martcheva, "Avian flu: Modeling and implications for control", *J. Biol. Syst.,* vol. 22, no. 1, pp. 151-175, 2014.
       http://dx.doi.org/10.1142/S0218339014500090

[13]   R.M. May, *Stability.* Princeton University Press: Princeton, NJ, 1974.

[14]   V.G. Matsenko, and V.N. Rubainovski, "Application of lyapunov direct method for analysing the age structure of biological population", *Zhvyscist Math. Fiz,* vol. 23, no. 2, pp. 320-332, 1983.

[15]   W. McCulloch Pitt, "A logical calculus of the ideas eminent in nervous activity", *Bull. Math. Biophys.,* vol. •••, pp. 105-111, 1943.

[16]   B.O. Oyelami, "δ-Controllability of impulsive systems and application to some physical and biological control", In: *Int. J. Differ.,* vol. 12. 2013no. 3, .

[17]   B.O. Oyelami, and S.O. Ale, "Impulsive differential equations and applications to some models: Theory and applications", *Lambert Academic Publisher Germany,* 2012. Available from: http://www.amazon.com/Impulsive-Differential-Equations-Applications-Models

[18]   B.O. Oyelami, *Studies in Impulsive Systems and Applications..* Lambert Academic Publisher Germany, 2012.

[19]   Kar, and Chaudhri, "Bionomic of multi-species fish and optimal harvesting", *Int. J. Math.Edu. Sci. Tech,* vol. 35, no. 4, 2002.

[20]   D. Tilman, R.M. May, C.L. Lehman, and M.A. Nowak, "Habitat destruction and the extinction debt", *Nature,* vol. 371, no. 6492, pp. 65-66, 1994.
       http://dx.doi.org/10.1038/371065a0

[21]   G. Deo Sadashiv, and V. Ragavendra, *Ordinary Differential Equations..* Tata McGraw-Hill: India, 1980.

[22]   G. Deo Sadashiv, V. Lakshimikantham, and V. Ragavendra, *Textbook of Ordinary Differential Equations.* Tata McGraw-Hill: India, 1997.

[23]   O. Ale Sam, and O. Oyelami Benjamin, "Mathematical modelling of the exploitations of biological resources in forestry and fishery", *Proceeding of National Mathematical Centre of the Workshop on Mathematical Modelling of Environmental Problems,* vol. 5, 2005pp. 6-34

# CHAPTER 9

# Numerical Solutions to Ordinary Differential Equations and Applications Using Maple

**Abstract:** Many complex nonlinear problems in science, economics, and engineering require computers to solve and simulate mathematical models describing them. Hence, it becomes extremely necessary to apply numerical methods to solve such problems. In this chapter, numerical methods for solving initial value problems are considered. Taylor series, Euler, Modified Euler's, Runge-Kutta, Adams-Bash forth-Moulton and Milne numerical methods are considered together with some Maple examples given. Numerical simulations are designed and implemented using Maple software for HIV/AIDS, Fitzhugh, and Fitzhugh-Nagumo, sickle cell anemia, zooplankton-fish, Gompertz tumor and neural firing models. The Explore facility in Maple 2022 is utilized to design sliders for investigating the behaviors of solutions of some ordinary differential equations subject to parameter change.

**Keywords:** Codes, Convergence, Fitzhugh, Fish models, Fitzhugh-Nagumo, Maple solution, Numerical, HIV/AIDS, Solutions, Stability, Sickle cell anemia, Tumor, Zooplankton.

## INTRODUCTION

In recent times, there are several emerging complex nonlinear ordinary differential equations (ODEs) in science and technology. The needs for computer-based solution to those ODEs problems are becoming increasingly important. Applications of numerical methods and development of numerical simulations are everywhere present in most research works in engineering, economics and life science these days [2,3,4,14]. The reason d'état is not far from the fact that the analytic or exact solutions to most non-linear ordinary differential equations cannot be easily being found even with the applications of symbolic programing.

It is interesting to note that, sometimes, the symbolic programs may take several hours, or even days to generate symbolic solutions to a problem. The computer printout of the solution may run into several pages of paper and the result may appear to be meaningless at a glance [12]. Hence the use of numerical methods in solving applied problems in science, economics and engineering is becoming popular in the recent times [3, 4, 5,8,12, 13].

**Benjamin Oyediran Oyelami**
**All rights reserved-© 2024 Bentham Science Publishers**

We note that, in the literature, much research has accumulated on numerical methods for solving ordinary differential equations. There are many physical systems with the governing equations, initial and boundary conditions, whose solution cannot be obtained with even some mathematical software in the market; yet the solution to the ODEs exist in some given interval. In this situation, numerical methods are often used to find the approximate solutions to the pig-headed ODE problems and numerical simulation now constitute the core of most researches on the behavior of the solution to the problems subject to parameter changes [2, 8, 12, 13].

Furthermore, the knowledge of numerical solution is very important to solve the differential equations [8]. We need efficient numerical methods in order to form algorithms for solving the problems. The algorithms must have desirable computation properties before being coded into computer programs. The programs are then implemented using higher level programming language to find the numerical solution to the differential equations. Maple software platform has given us the opportunity to gain understanding of the behavior of systems and discover laws underpinning them. It provides us with platforms to teach students how to find numerical solutions to ODEs and to develop numerical simulation to models [2, 5, 6, 9].

The numerical methods that we will consider are the Taylor series, Euler, Modified Euler's, Runge- Kutta, Adams-Bashforth, Adams-Bashforth-Moulton and Milne methods. It is worthy to note that each of these methods has some kind of complexities associated with it. These include: computer run time (time taken to run the program), computer memory (space occupied by the data generated from the numerical method); how fast the approximated solution tends to the analytic solution (convergence issue). The issue of consistency and stability of the methods are also paramount when considering numerical solutions [5,7,8,14]. Each of these computational properties will be discussed in our subsequent study on numerical solutions to ODEs.

Furthermore, in implementing numerical methods, a price must be paid which is associated with complexities in the numerical algorithms employed together with structuring programing language unitized to implement the numerical methods on the computer.

Finally, we will like to emphasize that numeric analysts would be interested in studying the numerical algorithms and the computational analytic concepts mentioned above, whereas other Scientists and Engineers will only be concerned with applying numerical methods to solve problems for as long as the solution generated from the numerical methods are accurate and satisfied the required computational properties.

**Numerical Solution to Initial Value Problems**

Let us consider an initial value problem (IVP):

$$\left.\begin{array}{l} \dfrac{dy(t)}{dx} = f(t, x(t)) \\[2mm] y(t_0) = y_0 \end{array}\right\} \tag{9.1}$$

where $f(t, x(t))$ is a continuous function in the closed interval $[a, b]$ and differentiable in the open interval $(a, b)$. Then by fundamental theorem of calculus the solution to the equation (9.1) is:

$$y(x) = y_0 + \int_{t_0}^{t} f(s, x(s))ds, t_0 = a, t \in [a, b] \tag{9.2}$$

Suppose the interval $[a, b]$ is partition into sub-intervals $[x_0, x_1), [x_1, x_2), [x_2, x_1), \dots [x_{k-1}, x_k)$ such that: $0 \le x_0 < x_1 < x_2 < \dots < x_k, x_{k+1} = x_k + h$, where h is constant.

We can find the Taylor series solution to the equation (9.1) as follows:

$$y(x) = y_0 + (x - x_0)y_0' + (x - x_0)^2 y_0'' + (x - x_0)^3 y_0'' + \dots + R_n \tag{9.3}$$

where $y_0 = y(x_0), y'_0 = y'(x_0), \dots, y_0^{(n)} = y^{(n)}(x_0)$. Therefore, the approximate solution to the differential equations can be generally be written as $y(x) = y(x_n) + \in_n = y_n + \in_n$, where $y_n$ is the approximate solution to $y(x)$ (exact solution) and $\in_n$ is the error of approximation.

Later on, we will discover that numerical methods differ by the value of $y(x)$ and the corresponding error value $\in_n, n = 1, 2, \dots$

## Taylor Series Solution Method

We consider the Taylor series method to find numerical solution of ordinary differential equations as follows:

$$y(x_{k+1}) = y_0 + (x_{k+1} - x_0)y_0' + (x_{k+1} - x_0)^2 y_0'' + (x_{k+1} - x_0)^3 y_0'' + ... + R_n \quad \textbf{(9.4)}$$

This can be rewritten as:

$$y_{k+1} = y_k + hy_k' + \frac{h^2}{2!} y_k'' + \frac{h^3}{3!} y_k'' + ... + \frac{h^k}{k!} y_k^{(k)/} + R_{k+1}$$

$$\textbf{(9.5)}$$

Where $y_{k+1} := y_k + h$. From iterative equation, in equation (9.5), we form different numeric algorithms for solving ODEs for given step size $h$ .

### Error, Convergence, Constituency and Stability for Numeric Methods

There are a lot of factors we need to take into consideration when finding numerical solutions to differential equations. We need to know whether the numerical method approximates the solution (convergence) and how well it approximates the solution (order of the method) and stability whether the errors of the method are damped out [5, 7, 8, 14].

### Definition 9.1

**Error**: the error of the numerical method is the difference between the analytic (exact) solution and the numerical solution obtained using the numerical method. The local truncation error of the method is the error committed by one step of the method.

**p- order**: a numerical method is said to be of p-order if for every iteration, the solution improves by p decimal places.

### Definition 9.2

**Convergence**: The solution of the numerical scheme is said to be convergent if the numerical solution converges towards to the exact solution of the ordinary differential equations. A numerical solution approaches the exact /analytic solution as step size $h \to 0$ .

**Consistency**: a numerical scheme is said to be consistent if its discrete operator (iterative scheme) converges towards the continuous operator (together with derivatives). A method is consistent if the order is greater than zero.

We note that consistency is a necessary condition for convergence but not sufficient.

**Definition 9.3**

**Stability**: a numerical method is said to be stable if the errors (truncation, round-off) errors *etc.* decays as the computation of the solution from one step to the next step. That is, the method does not allow magnification of error to propagate during computation process from one step to another.

**Numerical Methods**

Numerical methods for ordinary differential equations (ODEs) are methods used to find numerical approximation to the solution of ordinary differential equations; they are also known as numerical integration. ODEs numerical methods fall into two categories, linear multistep methods and the Kunge-Kutta methods. The methods can further be categorized as being implicit and explicit. The Euler method we will consider is an example of explicit method that is $y_{n+1}$ can be obtained from $y_n$ which is already known. The Backward Euler method is an implicit method which means we have to solve an equation to find $y_{n+1}$.

*Euler Method*

Euler's method is the simplest possible numerical solution to ordinary differential equations. The error of approximation is higher when the step size becomes large hence the method is often not employed to solve practical problems because the solution will not be accurate. The numerical algorithms for solving ODEs using the Euler's method is:

$$y_{n+1} = y_n + hy_n' = y_n + hf(x_n, y_n) = y_n + hf_n \qquad (9.6)$$

where $\dfrac{h^2}{2!} y_n''$ term is the neglected in the Taylor's series. The Trapezoidal method which is a second order method is given as:

$$y_{n+1} = y_n + \frac{h}{2}[y_n' + y_{n-1}']$$

**(9.7)**

## Backward Euler's Method

The backward Euler's method computes the value at $x + h$ with local truncation error proportional to $h^2$. Backward Euler's method is an implicit numerical method and always undershoots on the original curve.

The formula for the backward Euler's method is:

$$y'(x_1) = f(x_1, y_1)$$
$$y_1 = x_0 + hf(x_1, y_1)$$
$$y_2 = y_1 + hf(x_2, y_2)$$
$$y_3 = y_2 + hf(x_3, y_3)$$
$$y_{n+1} = y_n + hf(x_{n+1}, y_{n+1})$$

## Trapezium Rule Method

The Trapezoidal method is also known as Modified Euler's method. The method is the average of forward and backward Euler's methods and it is given as:

$$y(x+h) = y(x) + \frac{h}{2}\left[y'(x) + y'(x+h)\right]$$

**(9.8)**

That is $y_{n+1} = y_n + \frac{h}{2}\left[y_{n+1}' + y_n'\right]$

The local truncation error is propositional to $h^3$.

## Runge-Kutta of fourth –order Method

Let:

$$k_1 = hf(x_0, y_0)$$
$$k_2 = hf(x_0 + \frac{h}{2}, y_0 + \frac{k_1}{2})$$
$$k_3 = hf(x_0 + \frac{h}{2}, y_0 + \frac{k_2}{2})$$
$$k_4 = hf(x_0 + h, y_0 + k_3)$$

Then the algorithm for the fourth-order Runge-Kutta method is:

$$y = y_0 + \frac{1}{6}\left[k_1 + 2k_2 + 2k_3 + k_4\right] \tag{9.9}$$

The order of the error being $h^5$. The method is very accurate and no derivative is needed to compute the solution of the method.

## Multistep Methods

A predictor-corrector method is a set of two equations for $y_{n+1}$. The first equation which is often called a predictor is used to obtain first approximate to $y_{n+1}$, while the second equation is used to obtain second approximation to $y_{n+1}$ and this is called corrector.

## Modified Euler's Method

**Predictor**: $y_{n+1} + y_n + hy_n'$

**Corrector**: $y_{n+1} = y_n + \dfrac{h}{2}(y_{n+1}' + y_n')$

Now let $y_{n+1,1}' = f(x_{n+1}, y_{n+1,1})$ where $y_{n,1}$ is the predicted value of $y_n$.

## Adams-Bashforth-Moulton Method

**Predictor**: $y_{n+1} = y_n + \dfrac{h}{24}\left[55y_n' - 59y_{n-1}' + 37y_{n-2}' - 9y_{n-3}'\right]$

**Corrector**: $y_{n+1} = y_n + \dfrac{h}{24}\left[9y_{n+1,1}' + 19y_n' - 5y_{n-1}' + y_{n-2}'\right]$

## Milne's Method

**Predictor**: $y_{n+1,1} = y_{n-3} + \dfrac{4h}{3}\left[2y_n' - y_{n-1}' + 2y_{n-2}'\right]$

**Corrector**: $y_{n+1} = y_{n-1} + \dfrac{h}{3}\left[y_{n+1,1}' + 4y_n' + y_{n-1}'\right]$

Adams-Bashforth-Moulton and Milne's methods require information at $y_0, y_1, y_2$ and $y_3$ to start. The values for $y_0, y_1$ can be obtained from the initial conditions while $y_2$ and $y_3$ can be obtained from Runge-Kutta method. This kind of numerical method that requires multiple steps to solve ordinary differential equations is called a multistep method, whereas those that require a single step are called single step methods.

**Examples**

**Example 9.1**

Find the numerical solution to the following IVP using the Taylor's method:

$$\frac{dy}{dx} = x + y^2, y(0) = 0 \quad , \tag{9.10}$$

what is the value of $y(0.2)$ ?

**Solution**

$$\frac{dy}{dx} = x + y^2, y(0) = 0$$

Differentiating the above equation, we have:

$$\frac{d^2 y}{dx^2} = 1 + 2y\frac{dy}{dx}$$

$$\frac{d^3 y}{dx^3} = 2(\frac{d^2 y}{dx^2})^2 + 2y\frac{d^2 y}{dx^2}$$

$$\frac{d^4 y}{dx^4} = 4(\frac{dy}{dx})\frac{d^2 y}{dx^2} + 2\frac{dy}{dx}\frac{d^2 y}{dx^2} + 2y\frac{d^3 y}{dx^3}$$

Thus, when $x = 0, y = 0$

$$\frac{dy}{dx} = 0 + (0)^2 = 0$$

$$\frac{d^2 y}{dx^2} = 1 + 2(0)(0) = 1$$

$$\frac{d^3 y}{dx^3} = 2(0)^2 + 2(0)(1) = 0$$

$$\frac{d^4 y}{dx^4} = 4(0)(0) + 2(0)(1) + 2(0)(0) = 0$$

Therefore, the approximate solution is:

$$y = y_0 + xy_0' + \frac{x^2}{2!} y_0'' + \frac{x^3}{3!} y_0''' + \cdots$$

$$y = 0 + 0x + \frac{x^2}{2!}(1) + \frac{x^5}{5!}(1) = \frac{x^2}{2!} + \frac{x^5}{5!} =$$

$$= \frac{x^2}{2} + \frac{x^5}{20}$$

It implies that $y(0.2) \approx \dfrac{(0.2)^2}{2} + \dfrac{(0.2)^5}{20} = 0.020016$

## Example 9. 2

Find the numerical solution to the following IVP using the Taylor's method:

$$\frac{dy}{dx} = 3x + y^2, y(0) = 1 \tag{9.11}$$

What is the value of $y(0.5)$ ?

## Solution

$$x_0 = 0, y_0 = 1 (\text{give})$$

$$y'(x) = 3x + y^2(x), y_0' = 3(0) + y^2(0) = 0 + (1)^2 = 1$$

$$y'' = 3 + 2yy', y_0'' = 3 + 2y_0 y_0' = 3 + 2(1)(1) = 5$$

$$y''' = 2y'^2 + 2yy'', y_0''' = 2y_0'^2 + 2y_0 y_0'' = 2(1)^2 + 2(1)(5) = 12$$

$$y^{(iv)} = 2y'y''' + 2y'y'' + 2yy''', y_0^{(iv)} = 2y_0'y_0''' + 2y_0'y_0'' + 2y_0 y_0''' = 54$$

$$y = y_0 + (x - 0)y_0 + \frac{(x-0)^2}{2!} y_0'' + \frac{(x-0)^4}{4!} y_0^{(iv)}$$

$$y = 1 + x + \frac{5x^2}{2} + 2x^3 + \frac{9x^4}{4} + \cdots$$

Therefore:

$$y(0.5) \approx 1 + 0.5 + \frac{5(0.5)^2}{2} + 2(0.5)^3 + \frac{9(0.5)^4}{4} = 2.5156$$

## Example 9.3

Using the Euler's method to find the approximate solution to the following initial value problem:

$$\left.\begin{array}{l} \dfrac{dy}{dx} = 2x + 3y \\ \qquad y(1) = 1 \end{array}\right\} \tag{9.12}$$

## Solution

By Euler's algorithm: $y_{k+1} = y_k + hf(x_k, y_k) = y_k + hf_k$

Here in this case $f(x, y) = 2x + 3y, x_0 = 1, y_0 = 1$. Therefore, we have a numerical approximation represented in Table **9.1**.

**Table 9.1. Numerical of function using Euler's method.**

| $x$ | $y$ | $f(x,y) = 2x + 3y$ | $y + hf(x, y), h = 0.1$ |
|-----|------|--------------------|--------------------------|
| 1.0 | 1.00 | 5.0 | 1+0.1(5)=1.5 |
| 1.1 | 1.50 | 6.7 | 1.5+0.1(6.7)=2.17 |
| 1.2 | 2.17 | 9.6 | 2.17+0.1(9.6)=3.13 |
| 1.3 | 3.13 | 11.99 | 3.13+0.1(11.99)=4.33 |
| 1.4 | 4.33 | 38.77 | 4.33+0.1(38.77)=8.207 |
| 1.5 | 8.21 | 119.31 | 8.21+0.1(119.31)=20.14 |
| 1.6 | 20.14 | 63.62 | 20.14+0.1(63.62)=26.46 |

## MAPLE EXAMPLES

## Maple Example 9.1

### HIV/AIDS Model

Consider the HIV model:

$$\left.\begin{array}{c} \dfrac{dS(t)}{dt} = s + pS(t)\left(1 - \dfrac{S(t)}{S_{\max}}\right) - d_S S(t) - kv_0 S(t) \\[2mm] \dfrac{dU(t)}{dt} = kV(t)S(t) - \delta U(t) \\[2mm] \dfrac{dV(t)}{dt} = \delta N U(t) - cV(t) \end{array}\right\}$$

$$(9.13)$$

where $S(t)$ is the number of uninfected cells and $U(t)$ is the population of infected cells and $V(t)$ is the population of the HIV virus. We will simulate the model using Maple 2023 using the following parameters: $s = 0.01, p = 0.01, k = 0.005, \delta = 0.01, c = 0.05$, $S_0 = 2\times10^8, S_{\max} = 2\times10^6, U_0 = 3\times10^8$ and $V_0 = 4^7$.

We will find the numerical solution to the given HIV model using the above given parameters. The parameters were chosen arbitrarily but the actual value of the parameters can be obtained by experiment. The behavior of the three kinds of cells can be investigated from various scenarios.

We can investigate the behavior of the HIV model using Maple facility as follows:

**On the Maple worksheet Mode:**

**# to declare the maple facility for solving differential equations and plotting graphs type in:**

> *with(ODETools)* :

> *with(plots)* :

# # To calibrate the HIV model by inputting the parameters, type in:

> $s := 0.01$;

$s := 0.01$

> $p := 0.01$;

$p := 0.01$

> $k := 0.005$;

$k := 0.005$

> $b := 0.01$;

$b := 0.01$

>$S_{[max]} := 2.10^6$;

  $S_{max} := 2000000$

> $y[0] := 3 \cdot 10^8$;

$y_0 := 300000000$

> $v[0] := 4^7$;

$v_0 := 16384$

# # Enter in the HIV model equations

> $\qquad ode1 := diff(S(t), t) = s + p \cdot S(t) \cdot \left(1 - \dfrac{S(t)}{T[\max]}\right) - f \cdot S(t) - k \cdot v[0] \cdot S(t)$;

$\qquad ode1 := \dfrac{d}{dt} S(t) = 0.01 + 0.01\, S(t) \left(1 - \dfrac{1}{2000000}\, S(t)\right) - f S(t) - 81.920\, S(t)$

> $\qquad\qquad ode2 := diff(U(t), t) = k \cdot v[0] \cdot S(t) - b \cdot U(t)$;

$\qquad\qquad ode2 := \dfrac{d}{dt} U(t) = 81.920\, S(t) - 0.01\, U(t)$

> $ode3 := diff(V(t), t) = N \cdot b \cdot U(t) - c \cdot V(t);$

$$ode3 := \frac{d}{dt} V(t) = 0.01 \, N \, U(t) - c \, V(t)$$

# input more parameters

> $c := 0.05;$

$c := 0.05$

> $f := 0.3;$

$f := 0.3$

> $N := 1;$

$N := 1$

# To solve the problem, input the equations and the initial conditions

> $sol := \{ode1, ode2, ode3, S(0) = 1000, U(0) = 700, V(0) = 350\};$

$$sol := \left\{ S(0) = 1000, U(0) = 700, V(0) = 350, \frac{d}{dt} S(t) = 0.01 + 0.01 \, S(t) \left( 1 \right. \right.$$
$$\left. - \frac{1}{2000000} S(t) \right) - 82.220 \, S(t), \frac{d}{dt} U(t) = 81.920 \, S(t) - 0.01 \, U(t), \frac{d}{dt} V(t)$$
$$\left. = 0.01 \, U(t) - 0.05 \, V(t) \right\}$$

# Finding the numerical solution to the model, we invoke:

> $soln2 := dsolve(sol, numeric);$

$soln2 :=$ **proc**$(x\_rkf45)$ ... **end proc**

# To generate the table of values for the numerical solution invoke:

> **for** $j$ **from** 1 **to** 10 **do** $soln2(j)$ **end do**

$[t = 1., S(t) = 0.000121633811757597, U(t) = 1679.72204989651, V(t) = 349.280577847560]$

$[t = 2., S(t) = 0.000121646949602626, U(t) = 1663.01845129834, V(t) = 348.547886245208]$

$[t = 3., S(t) = 0.000121655294130018, U(t) = 1646.48105628786, V(t) = 347.688818300716]$

$[t = 4., S(t) = 0.000121646505206416, U(t) = 1630.10821112409, V(t) = 346.711150483796]$

$[t = 5., S(t) = 0.000121629361878961, U(t) = 1613.89827850010, V(t) = 345.622263951414]$

$[t = 6., S(t) = 0.000121624690520729, U(t) = 1597.84963738838, V(t) = 344.429163987116]$

$[t = 7., S(t) = 0.000121635185394149, U(t) = 1581.96068290875, V(t) = 343.138498490703]$

$[t = 8., S(t) = 0.000121652388213748, U(t) = 1566.22982616096, V(t) = 341.756575564572]$

$[t = 9., S(t) = 0.000121653424618842, U(t) = 1550.65549406902, V(t) = 340.289380240814]$

$$[t = 10., S(t) = 0.000121641291317122, U(t) = 1535.23612918375, V(t)$$
$$= 338.742590390979]$$

**# we can vary the initial condition to investigate the new behavior of cell in the model, in put**

> $sol4 := \{ode1, ode2, ode3, S(0) = 1000, U(0) = 500, V(0) = 320\};$

$$sol4 := \left\{ S(0) = 1000, U(0) = 500, V(0) = 320, \frac{d}{dt} S(t) = 0.01 + 0.01 S(t) \left( 1 \right. \right.$$
$$\left. - \frac{1}{2000000} S(t) \right) - 82.220 S(t), \frac{d}{dt} U(t) = 81.920 S(t) - 0.01 U(t), \frac{d}{dt} V(t)$$
$$\left. = 0.01 U(t) - 0.05 V(t) \right\}$$

> $soln3 := dsolve(sol4, numeric)$;

$soln3 := \mathbf{proc}(x\_rkf45)\ ...\ \mathbf{end\ proc}$

# # We generate the numerical solution using 4$^{th}$ and 5$^{th}$ order Runge-Kutta method

> **for** $j$ **from** 1 **to** 10 **do** $soln3(j)$ **end do**

$[t = 1., S(t) = 0.000121640516709076, U(t) = 1481.71208313999, V(t) = 318.802674650117]$

$[t = 2., S(t) = 0.000121653690545784, U(t) = 1466.97871663027, V(t) = 317.634700940589]$

$[t = 3., S(t) = 0.000121651258544514, U(t) = 1452.39194958218, V(t) = 316.380701151791]$

$[t = 4., S(t) = 0.000121637417454721, U(t) = 1437.95032330268, V(t) = 315.046293587738]$

$[t = 5., S(t) = 0.000121624045330204, U(t) = 1423.65239360526, V(t) = 313.636808387805]$

$[t = 6., S(t) = 0.000121629929559634, U(t) = 1409.49673066631, V(t) = 312.157301721539]$

$[t = 7., S(t) = 0.000121645251596472, U(t) = 1395.48191891753, V(t) = 310.612569289781]$

$[t = 8., S(t) = 0.000121655641960726, U(t) = 1381.60655688039, V(t) = 309.007159165953]$

$[t = 9., S(t) = 0.000121647603958052, U(t) = 1367.86925702057, V(t) = 307.345384009687]$

$[t = 10., S(t) = 0.000121631095586422, U(t) = 1354.26864558671, V(t)$
$= 305.631332683432]$

> ## ># To plot the graph input

$plots[odeplot](soln2, [[t, S(t), color = black], [t, U(t), color = blue], [t, V(t), color = red]], 0$
$.. 1.7, legend = [S(t), U(t), V(t)], labels = [t, ""])$

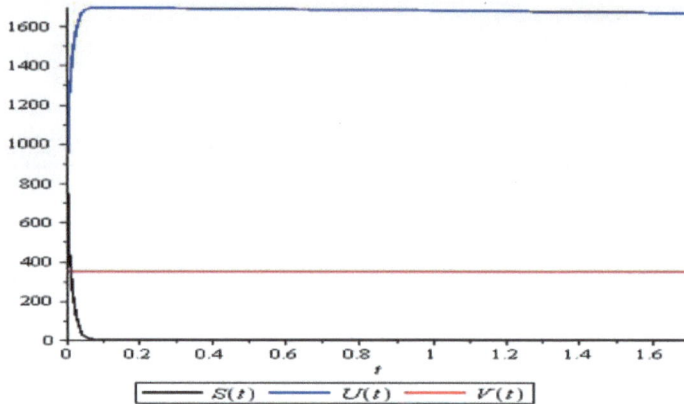

**Fig. (9.1).** Plot of graph for behavior of HIV model.

Fig. (**9.1**) is a graph for behavior of HIV models, from the graph, the number of uninfected cells is progressively decreasing to zero. The population of infected cells increases steadily and remains constant for some while, maybe because of the incubation period. The population of the HIV virus remains constant during the simulation period. We note that the choice of the parameters in the model determines the behavior of the HIV model.

# plot the graph of the solution to the model (Fig. 9.2)

> $plots[odeplot](soln3, [[t, S(t), color = red], [t, U(t), color = black], [t, V(t), color = blue]], 0$
   $.. 1.7, legend = [S(t), U(t), V(t)], labels = [t, ""])$

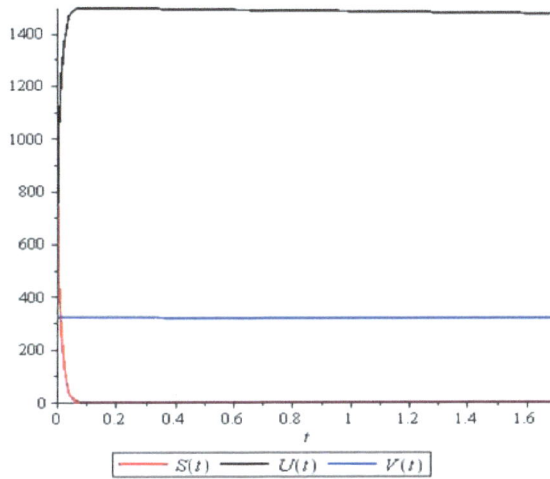

**Fig. (9.2).** The plot of the behavior of solution to the HIV model.

**# we can generate the numerical solution in array form as follows:**

**> dsol5 := dsolve(sol4, numeric, output=Array([0,0.25,0.5,0.75,1]));**

$dsol5 :=$

$$\left[\left[\left[\begin{bmatrix} t & S(t) & U(t) & V(t) \end{bmatrix}\right],\right.\right.$$

$[[[0., 1000., 500., 320.],$
$[0.250000000000000, 0.000122816722397827, 1492.85919848286, 319.618650063278],$
$[0.500000000000000, 0.000121640764742457, 1489.13420102518, 319.352563552957],$
$[0.750000000000000, 0.000121647745710870, 1485.41850325424, 319.080539469873],$
$[1., 0.000121640516709076, 1481.71208313999, 318.802674650117]]]]]$

**Maple Example 9.2**

*Sickle Cell Anemia Model*

Sickle cell anemia (SCA) is a genetic disorder commonly found among the black race of America, people of Africa and Mediterranean countries. The red blood cells of a SCA patient is sickled in shape which makes the deoxygenated sickle hemoglobin to cluster together. The clustered red blood cells block the free passage of carrying oxygen to the vital organ in the body. This often leads to physiological problems like stroke, heart and kidney failures, swollen of feats of the patient and damage of the brain [10, 11, 12].

Consider the model of oxygen contained in blood flow of sickle anemia [see [11] for the detail]:

$$
\left. \begin{array}{l}
\dot{x}_1(t) = -k_3 x_1(t) x_2(t) e^{(a+b)x} + k_3 x_3(t) e^{cx} + v_1 x_3(t) e^{cx} \\
\dot{x}_2(t) = -a v_0 k_1 x_1(t) e^{ax} - k_1 x_1(t) x_2(t) e^{(a+b)x} - k_1 x_1(t) x_3(t) e^{(a+b)x} + ac^2 x_3(t) e^{cx} \\
\dot{x}_3(t) = -k_1 x_1(t) x_3(t) e^{(a+b)x} + bc^2 x_3(t) e^{cx} - v_0 c x_{31}(t) e^{ax}
\end{array} \right\}
$$

$$(9.14)$$

Without loss of generality, we simulate the model for simple case when $a = b = c = 0$.

\# We start the numerical simulation for the sickle anemia as follows:

> *with(DETools)* :

> *with(plots)* :

> *ode1* := *diff*$(x(t), t)$ =$-k[3] \cdot x(t) \cdot z(t) \cdot e^{a \cdot x(t)} + k[3] \cdot z(t) \cdot e^{c \cdot z(t)} + c \cdot v[1] \cdot y(t) \cdot z(t) \cdot e^{c \cdot z(t)}$;

$$
ode1 := \frac{d}{dt} x(t) = -k_3 x(t) z(t) e^{a x(t)} + k_3 z(t) e^{c z(t)} + c v_1 y(t) z(t) e^{c z(t)}
$$

> *ode2* := *diff*$(y(t), t)$ =$-a \cdot v[1] \cdot y(t) \cdot e^{(a-b) \cdot x(t)} - k[1] \cdot x(t) \cdot z(t) \cdot e^{a \cdot x(t)} - k[1] \cdot x(t) \cdot y(t) \cdot z(t)$
$\cdot e^{(a+c-b) \cdot y(t)} + c^2 \cdot v[1] \cdot y(t) \cdot e^{(a-b) \cdot z(t)}$;

$$
ode2 := \frac{d}{dt} y(t) = -a v_1 y(t) e^{(a-b) x(t)} - k_1 x(t) z(t) e^{a x(t)} - k_1 x(t) y(t) z(t) e^{(a+c-b) y(t)}
$$
$$
+ c^2 v_1 y(t) e^{(a-b) z(t)}
$$

> *ode3* := *diff*$(y(t), t)$ =$-k[1] \cdot x(t) \cdot z(t) \cdot e^{a \cdot x(t)} - a \cdot c^2 \cdot z(t) + c \cdot v[1] \cdot x(t) \cdot e^{(a-b-c) \cdot z(t)}$;

$$ode3 := \frac{d}{dt} y(t) = -k_1 x(t) z(t) e^{a\, x(t)} - a\, c^2 z(t) + c\, v_1 x(t) e^{(a - b - c)\, z(t)}$$

> $a := 0; c := 0; b := 0;$

$a := 0$

$c := 0$

$b := 0$

> $k[1] := 4.23; v[1] := 1.25; k[3] := -0.254;$

$k_1 := 4.23$

$v_1 := 1.25$

$k_3 := -0.254$

> $ic := x(0) = 2000, y(0) = 3000., z(0) = 5000;$

$ic := x(0) = 2000, y(0) = 3000., z(0) = 5000$

> $ode := \{ode1, ode2, ode3, ic\};$

$$ode := \left\{ \frac{d}{dt} x(t) = 0.254\, x(t)\, z(t) - 0.254\, z(t), \frac{d}{dt} y(t) = -4.23\, x(t)\, z(t), \frac{d}{dt} y(t) = \right.$$
$$\left. -4.23\, x(t)\, z(t) - 4.23\, x(t)\, y(t)\, z(t), x(0) = 2000, y(0) = 3000., z(0) = 5000 \right\}$$

>
$sol := dsolve([ode1, ode2, ode3, x(0) = 12, y(0) = 34., z(0) = 50], numeric, range = 0..1, stiff$
$\quad = true);$

$sol := \mathbf{proc}(x\_rosenbrock\_dae) \; ... \; \mathbf{end\ proc}$

> **for** $j$ **from** 1 **to** 10 **do** $sol(j)$ **end do**

$[t = 1., x(t) = 12.0000000000000, y(t) = 34., z(t) = 0.]$

$[t = 2., x(t) = 12., y(t) = 34., z(t) = 0.]$

$[t = 3., x(t) = 12., y(t) = 34.0000000000000, z(t) = 0.]$

$[t = 4., x(t) = 12., y(t) = 34., z(t) = 0.]$

$[t = 5., x(t) = 12.0000000000000, y(t) = 34.0000000000000, z(t) = 0.]$

$[t = 6., x(t) = 12.0000000000000, y(t) = 34., z(t) = 0.]$

$[t = 7., x(t) = 12., y(t) = 34., z(t) = 0.]$

$[t = 8., x(t) = 12.0000000000000, y(t) = 34., z(t) = 0.]$

$[t = 9., x(t) = 12., y(t) = 34., z(t) = 0.]$

$[t = 10., x(t) = 12., y(t) = 34., z(t) = 0.]$

### > # we plot the graph by inputting in

$plots[odeplot](sol, [[t, x(t), color = black], [t, y(t), color = blue], [t, z(t), color = red]], 0..1,$
  $legend = [x(t), y(t), z(t)])$

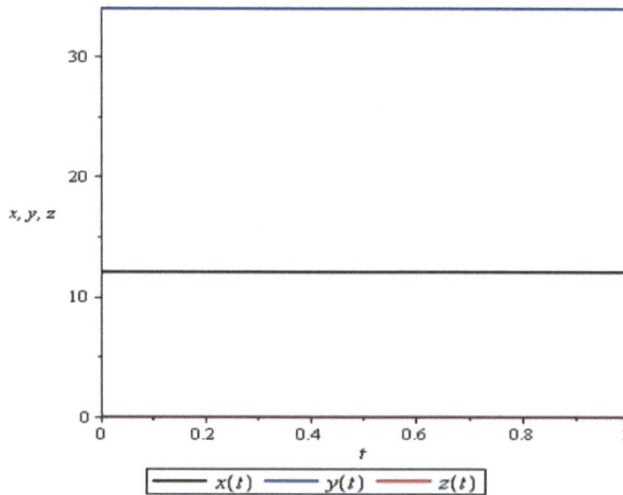

**Fig. (9.3).** Simulation of Sickle cell anemia model.

Fig. (**9.3**) is on simulation of Sickle cell anemia model; based on the choice of parameters $x(t), y(t)$ and $z(t)$ are constants throughout the simulation period. The simulation experiment can be replicated using different values of baseline data and parameters.

## Model for Oxygen Dissolved in Sickle Cell Anemia Patient Blood

Consider another model which is on the oxygen that dissolves in the hemoglobin of a sickle cell anemia patient when the patient uses oxygen enhancing drugs or supplements.

$$\frac{dV(t)}{dt} = cV(t) + dV(t)\exp(AV(t))$$

$$V(0) = V_0$$

**(9.15)**

Where $V(t)$ is the quantity of oxygen that dissolves in the hemoglobin of a sickle cell anemia patient at time $t$ [11].

#we calibrate the model by input in the parameters as follows:

> $c := 0.5; d := 1.23; A := 1.33;$

$c := 0.5$

$d := 1.23$

$A := 1.33$

#in put the model equation together with initial condition:

> $de2 := diff(V(t), t) = c \cdot V(t) + d \cdot V(t) \cdot \exp(-A \cdot V(t));$

$$de2 := \frac{d}{dt} V(t) = 0.5\, V(t) + 1.23\, V(t)\, e^{-1.33\, V(t)}$$

> $ic := V(0) = 10^5;$

$ic := V(0) = 100000$

# we find the analytic solution to the model using

> $dsolve(\{de2, ic\}, V(t))$

$$V(t) = RootOf\left( t - \left( \left( \int_{\_b}^{\_Z} \int_{\_a} \frac{100}{\_a \left( 50 + 123\,e^{-\frac{133}{100} - \_a} \right)}\, d\_a \right) + 100 \right) \left( \int_{\_b}^{100000} \int_{\_a} \frac{1}{\_a \left( 50 + 123\,e^{-\frac{133}{100} - \_a} \right)}\, d\_a \right) \right)$$

#from above, clearly the analytic solution depends on the roots of solution of integral equation which cannot be explicitly obtained. Therefore, we will obtain solution to the model by numerical method.

\# we the numeric solution to the model as follows:

> *with*(*Student*[*NumericalAnalysis*]) :

> *RungeKutta*(*de2, ic, t* = 20, *output* = *plot*)

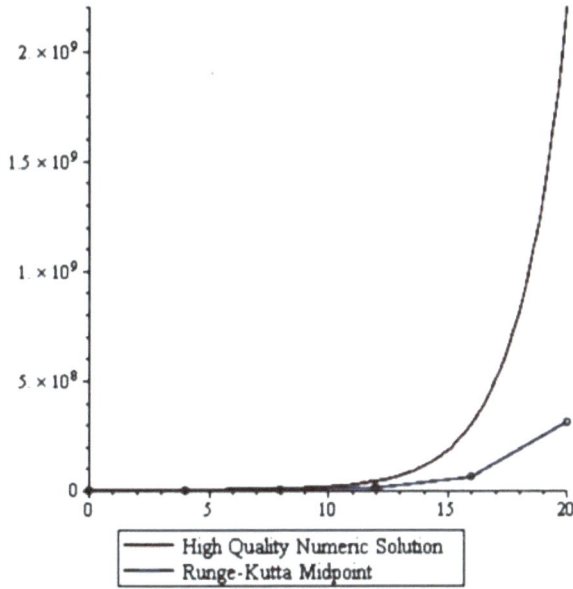

**Fig. (9.4).** Numerical solution to the model for finding Oxygen dissolved in sickle cell anemia patient blood.

Fig. (**9.4**) is on High quality numerical solution and Runge-Kutta midpoint solvers applied to compute oxygen that dissolved in blood of a sickle cell anemia patient. This is the case when a drug that enhances hemoglobin to absorb oxygen is taken.

We consider the situation when oxygen in the blood is depreciating continuously when no drug that enhances hemoglobin to absorb oxygen is being used, then c>0, the simulation result shown in Fig. (**9.5**). To achieve this, the simulation following maple codes was used:

```
> with(DETools) :
```

```
> with(plots) :
```

```
> with(Student[NumericalAnalysis]) :
```

```
> c := -2.5; d := 1.23; A := 2.33;
```

$c := -2.5$

$d := 1.23$

$A := 2.33$

$>$ *del* $:=$ *diff*$(V(t), t) = c \cdot V(t) + d \cdot V(t) \cdot \exp(-A \cdot V(t));$

$$del := \frac{d}{dt} V(t) = -2.5 V(t) + 1.23 V(t) \, e^{-2.33\,V(t)}$$

$>$ *ic* $:= V(0) = 10^5;$

$ic := V(0) = 100000$

$>$ *dsolve*$(\{del, ic\}, V(t))$

$$V(t) = RootOf\left( t - \left( \left[ \int_{\_b}^{\_Z} \frac{100}{\_a\left(-250 + 123\,e^{-\frac{233}{100}\_a}\right)} \, d\_a \right] + 100 \right. \right.$$
$$\left. \left. \left[ \int_{\_b}^{100000} \frac{1}{\_a\left(-250 + 123\,e^{-\frac{233}{100}\_a}\right)} \, d\_a \right] \right) \right)$$

$>$ *RungeKutta*$(del, ic, t = 20, output = plot)$

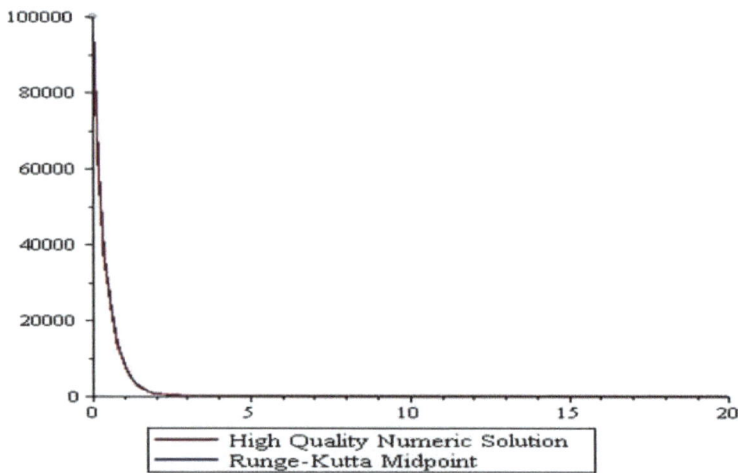

High Quality Numeric Solution
Runge-Kutta Midpoint

**Fig. (9.5).** Numerical solution to the model for finding Oxygen dissolved in sickle cell anaemia patient blood c<0.

## Maple Example 9.3

### *Zooplankton-Phytoplankton Population Model*

Zooplanktons are microscopic invertebrate animals that swim or drift in fresh water. They feed on bacteria and algae and other forms of particles in the water. Phytoplankton also known as microalgae are similar to plants since they contain chlorophyll for producing food. They are often found at the surface of ocean where sunlight is in abundance. Zooplanktons prey on phytoplankton and fish feeds on zooplanktons.

Phytoplankton is a major player in biogeochemical cycling of major elements like carbon, nitrogen, phosphorous and minor elements like iron, zinc and carbon dioxide in the ocean. The photosynthesis and nitrogen fixation attributes made it a major contributor to the global climate change.

Fish are a major source of protein and very rich in omega 3 fatty acid, one of food supplements good for human health especially enhancement of the development of memory in the brain.

Consider the zooplankton –phytoplankton population model as follows:

$$\left.\begin{array}{l} \dfrac{dP}{dt} = \beta \dfrac{Z^n}{H_Z^n + Z^n} P - cP^2 - \gamma \dfrac{P^n}{H_P^n + P^n} Z + D_p \\[3mm] \dfrac{dZ}{dt} = e\gamma \dfrac{P^n}{H_P^n + P^n} Z - \delta P^2 - \xi \dfrac{Z^n}{H_Z^n + Z^n} P + D_Z \end{array}\right\}$$

$$(9.16)$$

where $P$ and $Z$ are the populations of phytoplankton and zooplankton, respectively.

$\beta$ = maximum growth rate of the fish; $\gamma$ = growth rate of the zooplankton; $c$ = coefficient for competition for food for zooplankton and phytoplankton respectively; $e$ = prey assimilation coefficient of zooplankton and $\xi$ = feeding rate of zooplankton and $\delta$ = mortality rate.

$H_P^n$ and $H_Z^n$ are saturation constants for functional responses to nutrients by phytoplankton and zooplankton, respectively. $D_P$ and $D_Z$ are some relevant control parameters, here in this simulation, we will make use of circular functions [6].

#we apply the Maple to the zooplankton – phytoplankton model as follows:

```
> with(DETools) :
```

```
> with(plots) :
```

```
> with(Student[NumericalAnalysis]) :
```

# we enter the parameters

```
> n := 2; beta := 2.67;
```

$n := 2$

$\beta := 2.67$

```
> c := 0.5;
```

$c := 0.5$

```
> f := 1.34; e := 2.23; delta := 0.45; zeta := 1.23;
```

$f := 1.34$

$e := 2.23$

$\delta := 0.45$

$\zeta := 1.23$

# enter the control function

```
> w(t) := 1100· sin(t);
```

$w := t \rightarrow 1100 \sin(t)$

```
> u(t) := 1000· exp(-Pi·t)cos(t);
```

$u := t \rightarrow 1000\, e^{-\pi t} \cos(t)$

# Enter the equations

> $pop1 := \mathit{diff}(p(t), t) = \dfrac{\beta \cdot p(t) \cdot z(t)^n}{1 + z(t)^n} - c \cdot p(t)^2 - f \cdot \dfrac{z(t) \cdot p(t)^n}{1 + p(t)^n} + w(t);$

$pop1 := \dfrac{d}{dt} p(t) = \dfrac{2.67\, p(t)\, z(t)^2}{1 + z(t)^2} - 0.5\, p(t)^2 - \dfrac{1.34\, z(t)\, p(t)^2}{1 + p(t)^2} + 1100 \sin(t)$

> $pop2 := \dfrac{d}{dt} z(t) = \dfrac{e \cdot f \cdot z(t)\, p(t)^n}{1 + p(t)^n} - \mathrm{delta} \cdot p(t)^2 - \dfrac{\mathrm{zeta} \cdot p(t)\, z(t)^n}{1 + z(t)^n} + u(t);$

$pop2 := \dfrac{d}{dt} z(t) = \dfrac{2.9882\, z(t)\, p(t)^2}{1 + p(t)^2} - 0.45\, p(t)^2 - \dfrac{1.23\, p(t)\, z(t)^2}{1 + z(t)^2} + 1000\, e^{-\pi t} \cos(t)$

# Enter the model equations with the initial conditions

> $sol := \{pop1, pop2, p(0) = 1500, z(0) = 2500\};$

$sol := \left\{ \dfrac{d}{dt} p(t) = \dfrac{2.67\, p(t)\, z(t)^2}{1 + z(t)^2} - 0.5\, p(t)^2 - \dfrac{1.34\, z(t)\, p(t)^2}{1 + p(t)^2} + 1100 \sin(t), \dfrac{d}{dt} z(t) \right.$

$\qquad = \dfrac{2.9882\, z(t)\, p(t)^2}{1 + p(t)^2} - 0.45\, p(t)^2 - \dfrac{1.23\, p(t)\, z(t)^2}{1 + z(t)^2} + 1000\, e^{-\pi t} \cos(t), p(0) = 1500,$

$\qquad \left. z(0) = 2500 \right\}$

# invoking the numerical solver

> $soln := \mathit{dsolve}(sol, numeric)$

$soln := \mathbf{proc}(x\_rkf45) \; ... \; \mathbf{end\ proc}$

> $dsol5 := \mathit{dsolve}(sol, numeric, output = Array([0, 0.25, 0.5, 0.75, 1]))$

$dsol5 :=$

| $t$ | $p(t)$ | $z(t)$ |
|---|---|---|
| 0. | 1500. | 2500. |
| 0.250000000000000 | 0.371184485069144 | 1678.53956593776 |
| 0.500000000000000 | 0.499129232239408 | 1976.47658927639 |
| 0.750000000000000 | 0.557421954027257 | 2364.20047504374 |
| 1. | 0.566835641459682 | 2844.89056601254 |

# plotting the graphs

> $plots[odeplot](soln, [[t, p(t), color = black], [t, z(t), color = red]], 0 .. 1, legend = [p(t), z(t)])$

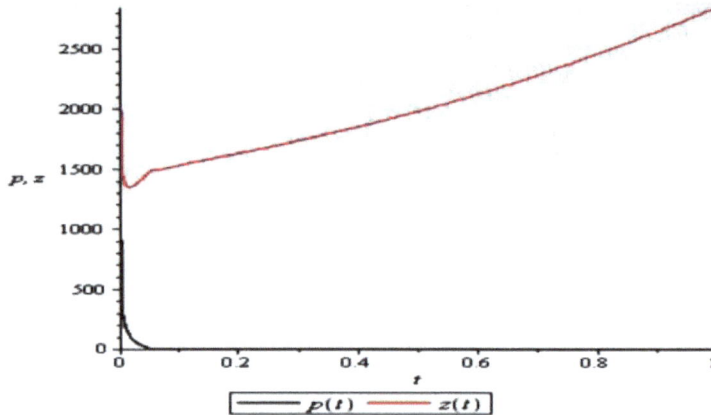

**Fig. (9.6).** Zooplankton and phytoplankton population.

Fig. (**9.6**) shows that the population of zooplankton increases, while phytopla-nkton population tend to cease, this is a persistent situation as it is often called in population dynamics.

> $$pop3 := diff(p(t), t) = \frac{\beta \cdot p(t) \cdot z(t)^4}{1 + z(t)^4} - c \cdot p(t)^2 - f \cdot \frac{z(t) \cdot p(t)^4}{1 + p(t)^4} + w(t);$$

$$pop3 := \frac{d}{dt} p(t) = \frac{2.67 \, p(t) \, z(t)^4}{1 + z(t)^4} - 0.5 \, p(t)^2 - \frac{1.34 \, z(t) \, p(t)^4}{1 + p(t)^4} + 1100 \sin(t)$$

> $$pop4 := \frac{d}{dt} z(t) = \frac{e \cdot f \cdot z(t) \, p(t)^4}{1 + p(t)^4} - delta \cdot p(t)^2 - \frac{zeta \cdot p(t) \, z(t)^4}{1 + z(t)^4} + u(t);$$

$$pop4 := \frac{d}{dt} z(t) = \frac{2.9882 \, z(t) \, p(t)^4}{1 + p(t)^4} - 0.45 \, p(t)^2 - \frac{1.23 \, p(t) \, z(t)^4}{1 + z(t)^4} + 1000 \, e^{-\pi t} \cos(t)$$

> $sol2 := \{pop3, pop4, p(0) = 1500, z(0) = 2500\}$

$$sol2 := \left\{ \frac{d}{dt} p(t) = \frac{2.67 p(t) z(t)^4}{1 + z(t)^4} - 0.5 p(t)^2 - \frac{1.34 z(t) p(t)^4}{1 + p(t)^4} + 1100 \sin(t), \frac{d}{dt} z(t) \right.$$

$$= \frac{2.9882 z(t) p(t)^4}{1 + p(t)^4} - 0.45 p(t)^2 - \frac{1.23 p(t) z(t)^4}{1 + z(t)^4} + 1000 e^{-\pi t} \cos(t), p(0) = 1500,$$

$$\left. z(0) = 2500 \right\}$$

> $soln2 := dsolve(sol2, numeric)$

$soln2 := \mathbf{proc}(x\_rkf45) \; ... \; \mathbf{end \; proc}$

>
$plots[odeplot](soln2, [[t, p(t), color = green], [t, z(t), color = red]], 0 .. 2, legend = [p(t), z(t)])$

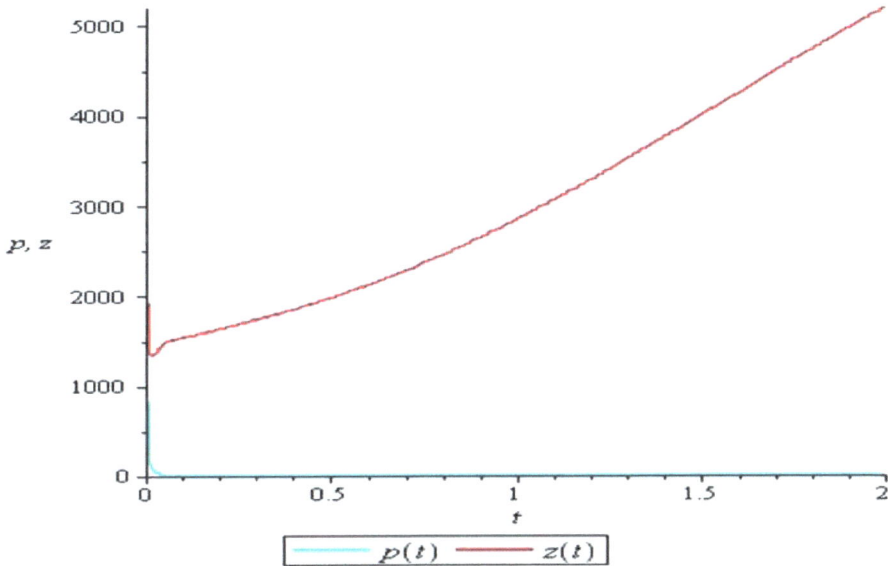

**Fig. (9.7).** Phytoplankton and zooplankton populations.

Fig. (**9.7**) shows that the population of zooplankton increases while phytoplankton population tends to cease, that is, it goes into extinction.

## Maple Example 9.4

### *Gompertz Tumor Growth Model*

Gompertz tumor growth model is the simplest form of cancer model for studying the growth of tumor in the body. In modern time, a variety of cancer models have evolved.

The Gompertz tumor growth model is as follows:

$$\frac{dx(t)}{dt} = kx(t)\ln(\frac{x_{max}}{x(t)})$$

$$x(0) = x_0$$

where $x =$ number of tumors is cells in the body at time $t$ and $x_{max} =$ maximum number of tumor cells present at time $t$ .

# we declare maple package and sub-package to be used:

> $with(DETools)$ :

> $with(plots)$ :

> $with(Student[NumericalAnalysis])$ :

# Input the parameters

> $k := .025;$

$k := 0.025$

> $s := 3 \cdot 10^5;$

# Enter the initial conditions,

$s := 300000$

> $ode1 := diff(y(t), t) = k \cdot y(t) \cdot \left( \ln\left( \frac{s}{y(t)} \right) \right);$

$$ode1 := \frac{d}{dt} y(t) = 0.025\, y(t)\, \ln\!\left( \frac{300000}{y(t)} \right)$$

> $IC := \left[ y(0) = 2 \cdot 10^2 \right];$

$$IC := [\, y(0) = 200 \,]$$

# Determine the analytic solution to the model:

> $ode2 := \left\{ ode1, y(0) = 2 \cdot 10^2 \right\};$

$$ode2 := \left\{ \frac{d}{dt} y(t) = 0.025\, y(t)\, \ln\!\left( \frac{300000}{y(t)} \right), y(0) = 200 \right\}$$

> $dsolve(ode2);$

$y(t)$

$$= 300000\ 200^{\mathrm{e}^{-\frac{1}{40} t}}\ 32^{-\mathrm{e}^{-\frac{1}{40} t}}$$

$$\mathrm{e}^{2\pi \left( 1\mathrm{e}^{-\frac{1}{40} t} {}_{\_Z1\sim} - 1\mathrm{e}^{-\frac{1}{40} t} {}_{\_Z3\sim} - 5\,\mathrm{I}\,\mathrm{e}^{-\frac{1}{40} t} {}_{\_Z4\sim} - 5\,\mathrm{I}\,\mathrm{e}^{-\frac{1}{40} t} {}_{\_Z2\sim} + 5\,\mathrm{I}\,\_Z2\sim \right) } {}_{3}{}^{-\mathrm{e}^{-\frac{1}{40} t}}$$

$$3125^{-\mathrm{e}^{-\frac{1}{40} t}}$$

# plot graphs:

> $gom := dsolve(ode2, numeric)$

$gom := \mathbf{proc}(x\_rkf45)\ \dots\ \mathbf{end\ proc}$

> **for** $j$ **from** 1 **to** 10 **do** $gom(j)$ **end do**

$[\, t = 1.,\ y(t) = 239.578571362460 \,]$

$[\, t = 2.,\ y(t) = 285.712859190360 \,]$

$[\, t = 3.,\ y(t) = 339.252655630399 \,]$

$[\, t = 4.,\ y(t) = 401.120629798889 \,]$

$[t = 5., y(t) = 472.313647128272]$

$[t = 6., y(t) = 553.903498482425]$

$[t = 7., y(t) = 647.036948413872]$

$[t = 8., y(t) = 752.935170086115]$

$[t = 9., y(t) = 872.892513774644]$

$[t = 10., y(t) = 1008.27452406737]$

> $plots[odeplot](gom, [[t, y(t), color=black]], 0 .. 20)$

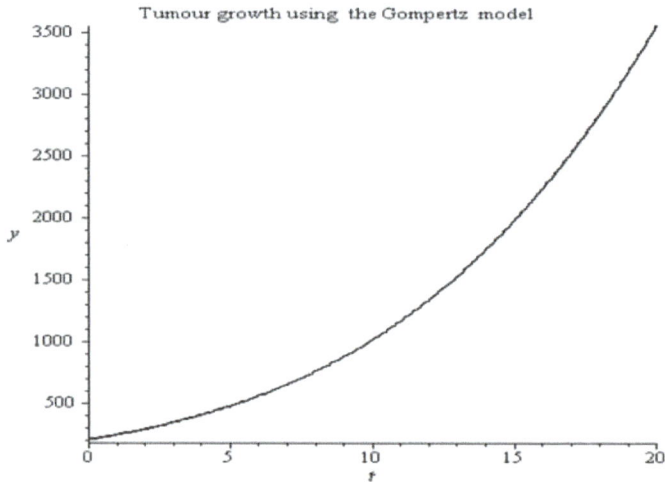

**Fig. (9.8).** Tumor growth using Gompertz model.

Fig. **(9.8)** shows tumor grows exponentially using the Gompertz model. Uncontrollable growth is very characteristic of benign and malignant cancer.

# Investigate another scenario with change of parameters:

> $k := 0.25; x\_max := 5000;$

$k := 0.25$

$x\_max := 5000$

$>$ $del := diff(z(t), t) = k \cdot z(t) \cdot \ln\left(\dfrac{5000}{z(t)}\right);$

$$del := \frac{d}{dt} z(t) = 0.25 \, z(t) \, \ln\left(\frac{5000}{z(t)}\right)$$

$>$ $ic := z(0) = 1000;$

$$ic := z(0) = 1000$$

$>$ $dsolve(\{del, ic\}, z(t))$

$$z(t) = 5000 \; 1000^{e^{-\frac{1}{4}t}} \; e^{-c^{-\frac{1}{4}t}(8 \, I \pi \_Z2\sim + 6 \, I \pi \_Z3\sim - 6 \, I \pi \_Z1\sim + 3 \ln(2))} \; 625^{-e^{-\frac{1}{4}t}}$$

\# We design a slider using the Explore facility in the Maple software. The slider helps us to see how a variable changes with parameters. How are the variables sensitive to parameter changes? This is the question a designed slider tends to answer (Fig. **9.9**).

$>$ $with(Student[NumericalAnalysis]):$

$>$ $RungeKutta(del, ic, t = 50, output = plot)$

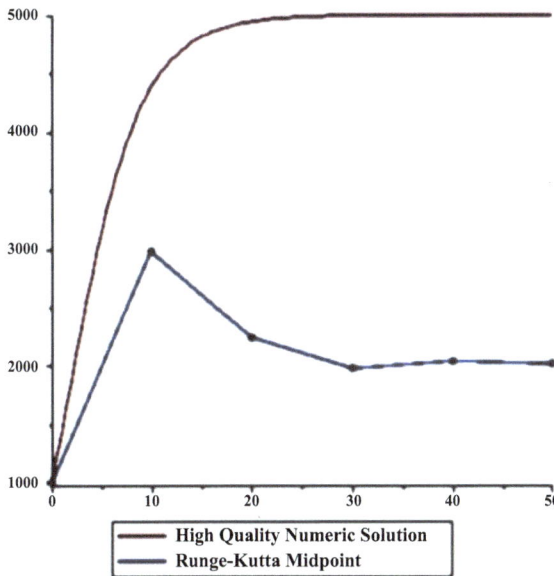

**Fig. (9.9).** Numerical solution to Gomperz model.

\# To determine the behavior of Gompertz model using sliders designed using the RungeKutta solver (Fig. **9.10**).

> *Explore(RungeKutta(de1, ic, t = a, output = plot))*

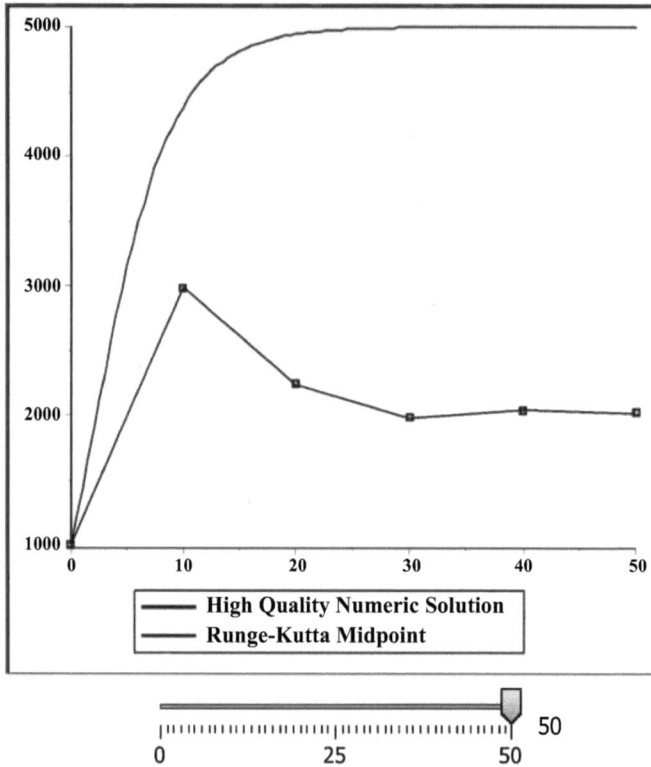

**Fig. (9.10).** Investigating the behavior of solution of the Gompertz model using a slider.

## Maple Example 9.5

>
$$with(Student[NumericalAnalysis]) :$$
$$Euler(diff(z(t), t) = 1/(2*t - 3*z(t) + 5), z(0) = 1, t = 1, output = plot);$$

Fig. (**9.11**) shows the numerical solution to the Maple example 9.5 found using Euler numerical approximation method and the result is compared with maple inbuilt high quality numeric solution.

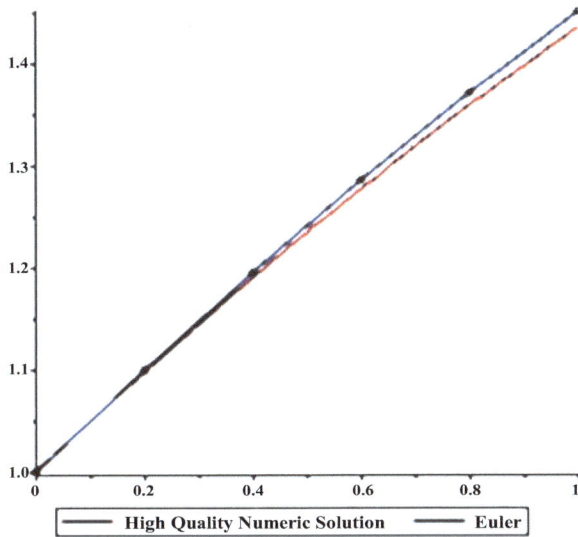

**Fig. (9.11).** Numerical solution to the Maple Maple example 9.5.

## > **Maple Example 9.6 (Fig. 9.12)**

$with(Student[NumericalAnalysis])$ :
$RungeKutta(diff(z(t), t) = 1/(2*t-3*z(t) + 5), z(0) = 1, t = 1, output = plot);$

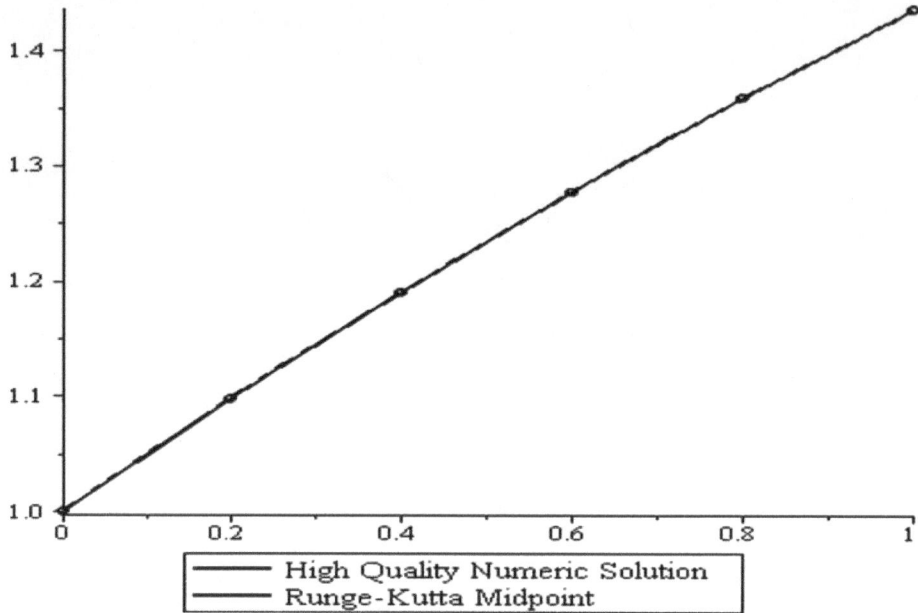

**Fig. (9.12).** Numerical solution to the Maple example 9.6 using Runge-Kutta method.

## Maple Example 9.7 (Fig. **9.13**)

>

$with(Student[NumericalAnalysis])$ :

$RungeKutta\left( diff(z(t), t) = 1/\left(2*t^2 - 3*z(t) + 5\right), z(0) = \dfrac{1}{2}, t = 1, output = plot \right)$

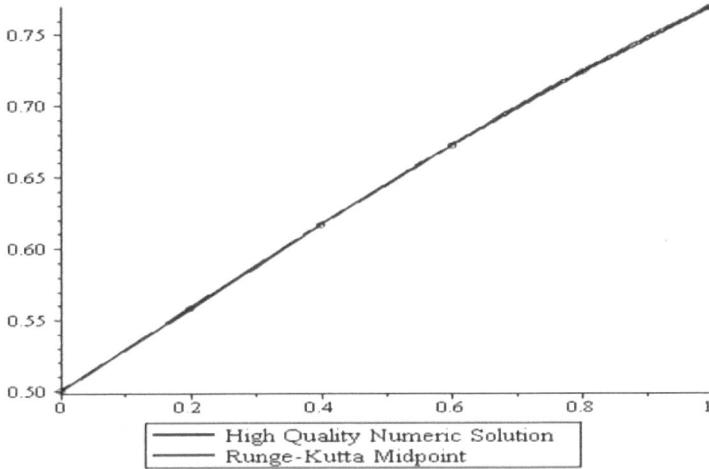

**Fig. (9.13).** Numerical solution to the Maple example 9.7.

## Maple Example 9.8

Let the simple speed of a rocket (Fig. **9.14**) be given as:

$$\frac{dx(t)}{dt} = \frac{1}{200}\ln(\frac{1}{5000}e^{-0.02t})$$

$$x(0) = x_0 = 100ms^{-1}$$

Where $x(t)$ is the vertical distance of the rocket above launch position. To solve the rocket problem and plot the graph of numerical solution to the problem, we make use of Runge-Kutta method in Maple as follows:

> 
    $with(Student[NumericalAnalysis]):$

    $RungeKutta\left(\frac{d}{dt}x(t) = \frac{1}{200}\ln\left(\frac{1}{5000}e^{-0.02\,t}\right), x(0) = 100, t = 1, output = plot\right)$

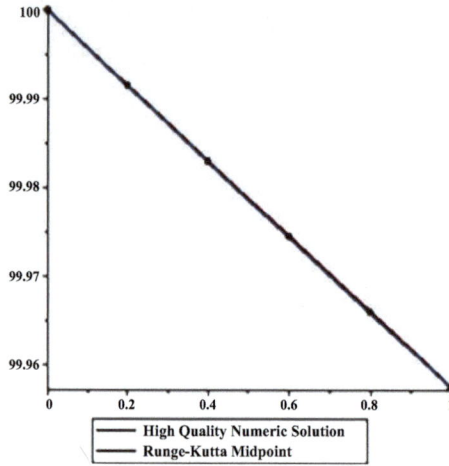

**Fig. (9.14).** Speed of rocket with time in seconds.

> $$speed := \frac{d}{dt} x(t) = -\frac{3}{200} \ln(2) - \frac{1}{50} \ln(5) + \frac{1}{200} \ln\left(e^{-\frac{1}{50} t}\right)$$

$$speed := \frac{d}{dt} x(t) = -\frac{3}{200} \ln(2) - \frac{1}{50} \ln(5) + \frac{1}{200} \ln\left(e^{-\frac{1}{50} t}\right)$$

> $RungeKutta(speed, x(0) = 10000, t = 1, output = plot)$

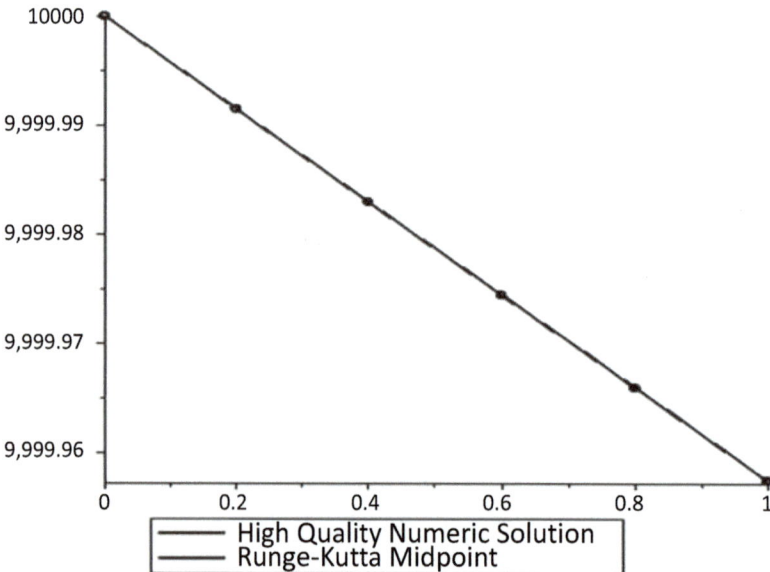

**Fig. (9.15).** Speed of rocket with time in seconds.

Figs. (**9.15** and **9.16**)approached same rocket problem using different numerical methods in the maple and obtained similar results.

## Maple Example 9.9

### *Plant Growth Model:*

> 
$$plantgrowth := \frac{d}{dt} w(t) = \frac{1}{200} (0.03\, t + 0.02)\, w(t);$$

$$plantgrowth := \frac{d}{dt} w(t) = \frac{1}{200} (0.03\, t + 0.02)\, w(t)$$

> $RungeKutta(plantgrowth, w(0) = 10000, t = 1, output = plot)$

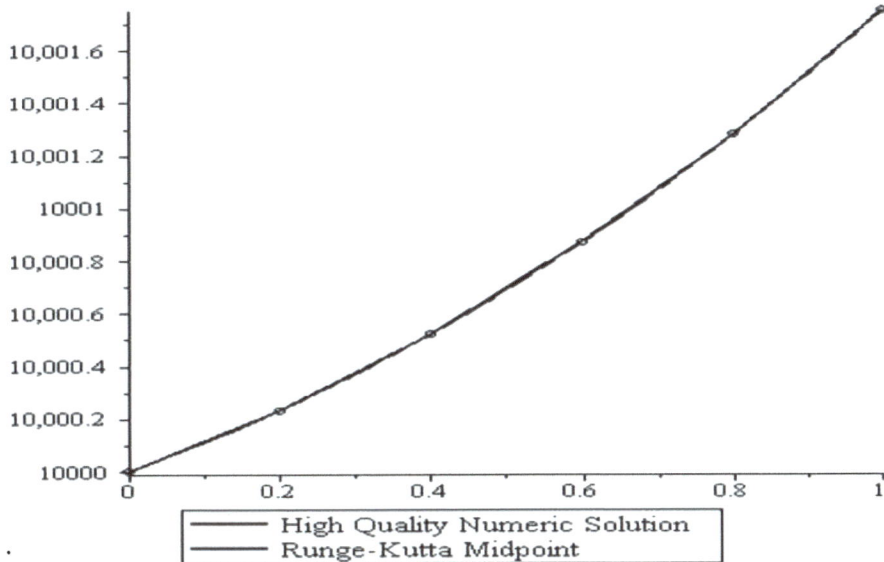

**Fig. (9.16).** Numerical solution to the simple rocket problem.

> $dsolve(plantgrowth, [w(t)], [[w(0) = 100]]));$

## Maple Example 9.10

We make use of the InitialValueProblem facility in Maple to solve the following initial value problem:

$$\frac{dy(t)}{dt} = y(t) + \frac{2y^2(t)}{1+y^2(t)} + 1 - t^2$$

$$y(0) = 0.5$$

> *restart*

> *with(Student[NumericalAnalysis])* :

>
$$DE1 := \frac{d}{dt} y(t) = y(t) + \frac{2 \cdot y^2(t)}{1 + y^2(t)} - t^2 + 1 :$$

#### #we evaluate the solution of the problem for $t = 2$

> *InitialValueProblem(DE1, y(0) = 0.5, t = 2)*

9.736

### # To compute the error of approximation, we use

> *InitialValueProblem(DE1, y(0) = 1, t = 3, output = Error)*

24.88

>
$$DE2 := \frac{d}{dt} y(t) = 1 - \cos(t) :$$

>
$$DE3 := \frac{d}{dt} y(t) = y(t) + \frac{2 \cdot y^2(t)}{1 + y^2(t)} - t^2 + \frac{t^3}{9} :$$

># we find the numerical solution to the equation DE2 when $y(1) = 3.10$ using Runge-Kutta and Taylor's series methods and compare their error of computations.
*InitialValueProblem(DE2, y(1) = 3.10, t = 5, method = rungekutta, submethod = rkf, comparewith = [[taylor, 2]], output = information, digits = 3)*

Table **9.2** displays Numerical solution using R-K-F and Taylor's methods and error of approximation of solution for Maple Example 9.10.

## Table 9.2: Numerical solution using R-K-F and Taylor's methods and error of approximation of solution.

| $t$ | [Maple's numeric solution] | [R-K-F] | [Error] | [2nd-Ord. Taylor] | [Error] |
|---|---|---|---|---|---|
| 1. | 3.10 | 3.10 | 0. | 3.10 | 0. |
| 1.80 | 3.77 | 3.77 | 0.00238 | 3.74 | 0.03 |
| 2.60 | 5.03 | 5.03 | 0.00403 | 5.03 | 0. |
| 3.40 | 6.60 | 6.60 | 0.00299 | 6.68 | 0.08 |
| 4.20 | 8.01 | 8.01 | 0.00305 | 8.17 | 0.16 |
| 5. | 8.90 | 8.90 | 0.000395 | 9.09 | 0.19 |

# From above, it is clear that RKF has the least error compared to the Taylor's series method. Therefore, RKF method is very accurate and it is a third order method compared to the Taylor's method and the Euler's methods which are of first order.

>

*InitialValueProblem(DE2, y(1) = 3.10, t = 5, method = rungekutta, submethod = rkf, comparewith = [[taylor, 1], [taylor, 2]], output = plot)*

Figs. (**9.17** and **9.18**) are numerical solutions using Runge-Kutta Fehlberg method compared with Taylor series methods and Euler method. The results obtained are compared the inbuilt high quality numeric solvers

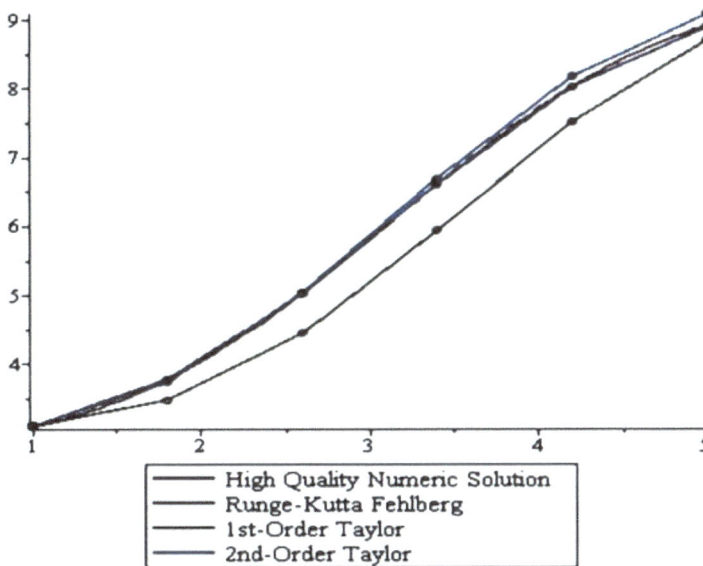

**Fig. (9.17).** Numerical solution using Runge -Kutta- Fehberg and Taylor's methods.

## Maple Example 9.11

>

*with*( *Student*[ *NumericalAnalysis* ]) :
*Euler*( *diff*($z(t), t$) = $1/(2*t-3*z(t)+5)$, $z(0) = 1, t = 1, output = plot$);

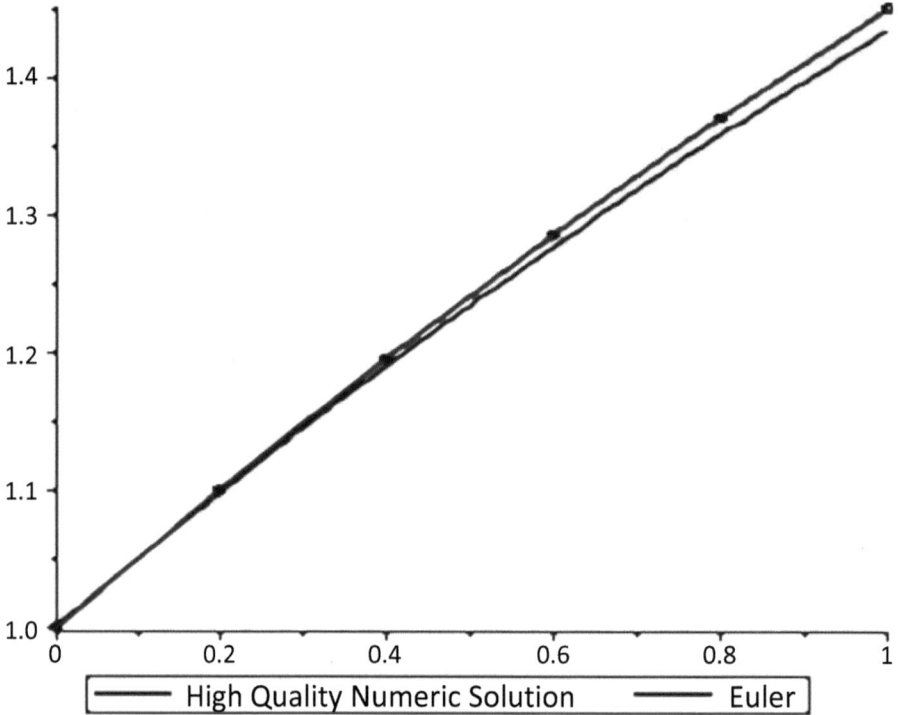

**Fig. (9.18).** Numerical solution using inbuilt numeric method compare with Euler's method.

Figs. **9.19** and **9.20** are on Numerical solutions using standard inbuilt numeric method, Runge-Kutta compared it with lower order method such as Euler approximation method.

## Maple Example 9.12

>

*with*( *Student*[ *NumericalAnalysis* ]) :
*RungeKutta*( *diff*($z(t), t$) = $1/(2*t-3*z(t)+5)$, $z(0) = 1, t = 1, output = plot$)

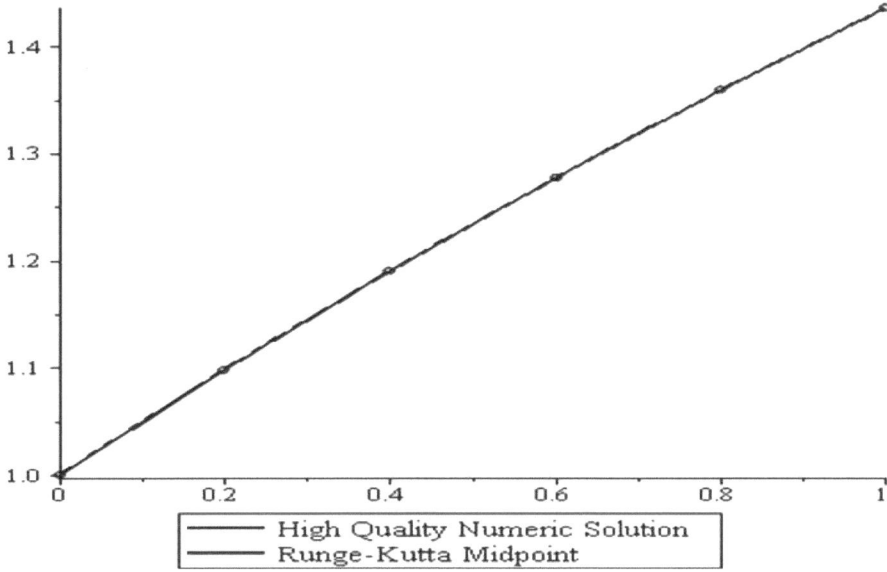

**Fig. (9.19).** Numerical solution using inbuilt numeric method compare with Runge-Kutta method.

>

$with(Student[ NumericalAnalysis ])$ :
$Euler(diff(z(t), t) = 1 / (2*t - 3*z(t) + 5), z(0) = 1, t = 1, output = plot)$;

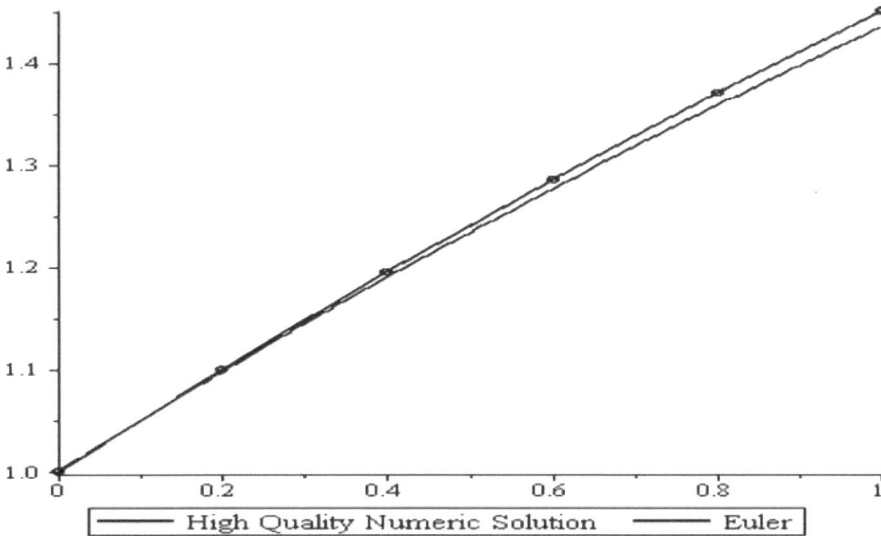

**Fig. (9.20).** Numerical solution using inbuilt numeric method compare with Euler's method.

## Maple Example 9.13

### *Fitzhugh-Nagumo Model of Neuron Firing*

Fitzhugh-Nagumo model is a simplification of the model due to Hodgkin and Huxley is the basis of neural system. The model relates the potential of the cell membrane, the permeability of the membrane and the applied current. This model is given by the equation:

$$\left.\begin{array}{c} \dot{v} = v(a-v)(v-1) - w + I \\ \dot{w} = \beta v - \gamma w \end{array}\right\}$$

where $v$ represents the membrane potential, $w$ is the restoring force, I the applied current, $a, \beta$ and $\gamma$ are constants.

# enter the Fitzhugh-Nagumo model together with the initial conditions:

$ode := \{diff(v(t), t) = v(t) \cdot (a - v(t)) \cdot (v(t) - 1) - w(t) + z, diff(w(t), t) = b \cdot v(t) - c$
$\cdot w(t), v(0) = 2 \cdot 10^{-9}, w(0) = 1.5 \cdot 10^{-9}\};$

$$ode := \left\{\frac{d}{dt} v(t) = v(t) (21 - v(t)) (v(t) - 1) - w(t) + 5, \frac{d}{dt} w(t) = -0.02 v(t)\right.$$

$$\left. - 0.234 w(t), v(0) = \frac{1}{500000000}, w(0) = 1.500000000 \ 10^{-9}\right\}$$

>     $plots[odeplot](soln, [[t, v(t), color=black], [t, w(t), color=blue]], 0..50, legend=[ v(t),$
      $w(t)])$

>For simulation for:

Fig. (**9.21**) is a solution to Fitzhugh-Nagumo neuron firing model; the membrane potential traces out ogive curve path; whereas, the restoring force traces a path that is literally an inverted image about an imaginary axis horizontally passing through w=0.2.

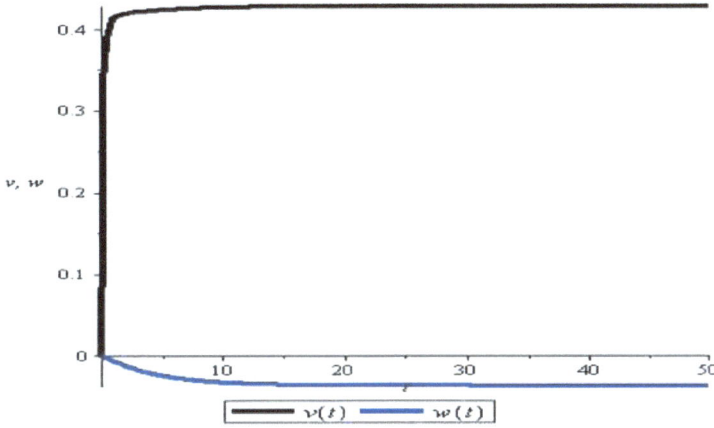

**Fig. (9.21a).** Solution of Fitzhugh-Nagumo neuron firing model.

$b := 0.02$ ,$a := 21$ ,$c := 0.234$ ,$v(0) = \dfrac{1}{500000000}$, $w(0) = 1.500000000 \, 10^{-9}$

We have the curve

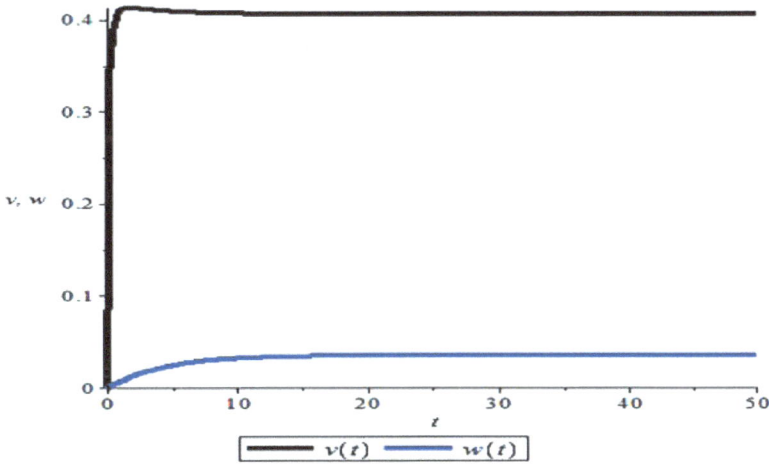

**Fig. (9.21bs).** Solution of Fitzhugh-Nagumo model2.

## Maple Example 9.14

### *Fitzhugh Model*

We consider the Fitzhugh model (Fig. **9.22**) as a Van der Pol model with some term added (inducing asymmetry). It also includes a model of arrhythmia:

$$\frac{dx_i}{dt} = y_i - 1.5 + \left(x_i - \left(\frac{x_i^3}{3}\right)\right) - A(\sin(0.2t)) \quad \frac{dy_i}{dt} = -\left(\frac{1}{9}\right)(x_i - 0.467 + 0.8(y_i - 1.5))$$

*with(DEtools)* :

> *with(plots)* :

> 
$$ode1 := diff(x(t), t) = y(t) - 1.5 + \left(x(t) - \frac{x^3(t)}{3}\right) - A \cdot \sin(0.2 \cdot t);$$

$$ode1 := \frac{d}{dt} x(t) = y(t) - 1.5 + x(t) - \frac{1}{3} x(t)^3 - 1.25 \sin(0.2 t)$$

> 
$$ode2 := diff(y(t), t) = -\frac{1}{9} \cdot (x(t) - 0.467 + 0.8 \cdot (y(t) - 1.5));$$

$$ode2 := \frac{d}{dt} y(t) = -\frac{1}{9} x(t) + 0.1852222222 - 0.08888888889 y(t)$$

> $A := 1.25;$

$A := 1.25$

> $x0 := 1;$

$x0 := 1$

> $y0 := 2;$

$y0 := 2$

> $sol := dsolve([ode1, ode2, x(0) = x0, y(0) = y0], numeric, output = listprocedure);$

$sol := [t = \mathbf{proc}(t) \ ... \ \mathbf{end \ proc}, x(t) = \mathbf{proc}(t) \ ... \ \mathbf{end \ proc}, y(t) = \mathbf{proc}(t) \ ... \ \mathbf{end \ proc}]$

> $X\_ans := rhs(sol[2]); \quad Y\_ans := rhs(sol[3]);$

$X\_ans := \mathbf{proc}(t) \ ... \ \mathbf{end \ proc}$

$Y\_ans := \mathbf{proc}(t) \ ... \ \mathbf{end \ proc}$

> $plot([X\_ans(t), Y\_ans(t), t = -100..100]);$

**Numeric solution of Fitzhugh model**

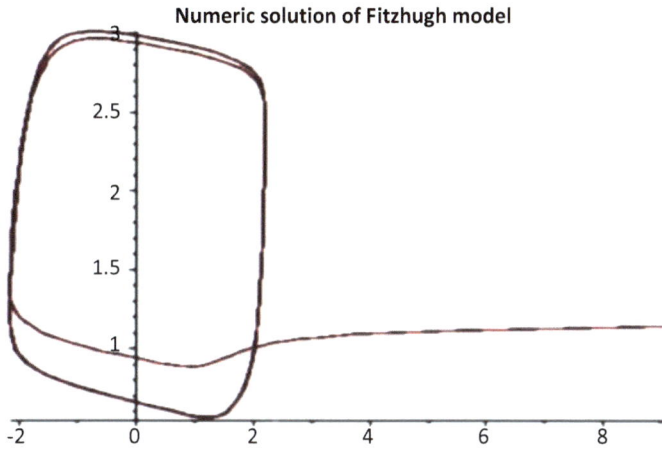

**Fig. (9.22).** Fitzhugh model.

> $A := 1.25;$  .

$A := 1.25$

> $ode1 := diff(x(t), t) = y(t) - 1.5 + \left( x(t) - \dfrac{x^3(t)}{3} \right) - A \cdot \sin(0.2 \cdot t);$

$$ode1 := \frac{d}{dt} x(t) = y(t) - 1.5 + x(t) - \frac{1}{3} x(t)^3 - 1.25 \sin(0.2\,t)$$

> $ode2 := diff(y(t), t) = -\dfrac{1}{9} \cdot (x(t) - 0.467 + 0.8 \cdot (y(t) - 1.5));$

$$ode2 := \frac{d}{dt} y(t) = -\frac{1}{9} x(t) + 0.1852222222 - 0.08888888889\, y(t)$$

> $ode := [ode1, ode2];$

$$ode := \left[ \frac{d}{dt} x(t) = y(t) - 1.5 + x(t) - \frac{1}{3} x(t)^3 - 1.25 \sin(0.2\,t),\ \frac{d}{dt} y(t) = -\frac{1}{9} x(t) \right.$$
$$\left. + 0.1852222222 - 0.08888888889\, y(t) \right]$$

> $ics := x(0) = 1, y(0) = 2;$

$ics := x(0) = 1, y(0) = 2$

> $systeode := \{ode1, ode2, x(0) = 1, y(0) = 2\};$

$systeode := \{ode1, ode2, x(0) = 1, y(0) = 2\}$

> $systeode : \{ode1, ode2, x(0) = 1, y(0) = 2\};$

$$\left\{ \frac{d}{dt} x(t) = y(t) - 1.5 + x(t) - \frac{1}{3} x(t)^3 - 1.25 \sin(0.2\,t), \frac{d}{dt} y(t) = -\frac{1}{9} x(t) \right.$$
$$\left. + 0.1852222222 - 0.08888888889 y(t), x(0) = 1, y(0) = 2 \right\}$$

**Fig. (9.23).** is plot of solution to Fitzhugh model  using interactive dynamic system facility in Maple.

> $dsolve['{:}\text{-}interactive']( \mathbf{(14)} )$

soll := dsolve([diff(x(t),t) = y(t)-1.5+x(t)-1/3*x(t)^3-1.25*sin(.2*t), diff(y(t),t) = -1/9*x(t)+.1852222222-.8888888889e-1*y(t), x(0) = 1, y(0) = 2], numeric);

soll (0.);

Plots [odeplot] (soll, 0...10, color = red);

> $dsolve['{:}\text{-}interactive']( \mathbf{(14)} )$

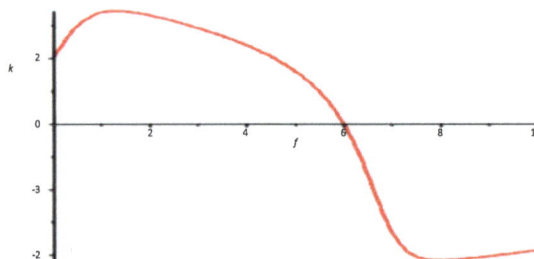

**Fig. (9.23).** Fitzhugh model solution plot using interactive dynamic system facility in Maple.

## Maple Example 9.15

> $with(Student[NumericalAnalysis])$ :

>
$$f := (x, y) \rightarrow \frac{1}{\left(3 \cdot x - 6 \cdot y^2 - y + 5\right)};$$

$$f := (x, y) \rightarrow \frac{1}{3x - 6y^2 - y + 5}$$

> $x[0] := 0; x[-1] := 0;$

$x_0 := 0$

$x_{-1} := 0$

> $y[0] := 1; y[-1] := 0;$

$y_0 := 1$

$y_{-1} := 0$

> $h := 0.1;$

$h := 0.1$

> **for** $k$ **from** $0$ **to** $10$ **do** $x[k] := k \cdot h : y[k] := y[k-1] + h \cdot f(x[k-1], y[k-1]);$**end do**

In Table **9.3**, we generate numerical solution to the Maple Example 9.14 using Euler approximating method.

**Table 9.3. Euler's numerical approximation to solution of problem.**

| $x$ | $y$ |
|---|---|
| $x_0 := 0.$ | |
| $x_1 := 0.1$ | $y_1 := 0.04009000321$ |
| $x_2 := 0.2$ | $y_2 := 0.05913665452$ |

(Table 9.3) cont.....

| | |
|---|---|
| $x_3 := 0.3$ | $y_3 := 0.07725298880$ |
| $x_4 := 0.4$ | $y_4 := 0.09453328194$ |
| $x_5 := 0.5$ | $y_5 := 0.1110571617$ |
| $x_6 := 0.6$ | $y_6 := 0.1268926232$ |
| $x_7 := 0.7$ | $y_7 := 0.1420982868$ |
| $x_8 := 0.8$ | $y_8 := 0.1567251194$ |
| $x_9 := 0.9$ | $y_9 := 0.1708177678$ |
| $x_{10} := 1.0$ | $y_{10} := 0.1844156063$ |

> $restart\ ;\ with(DETools)\ :$

> $with(plots)\ :$

> $$ode1 := diff(y(x), x) = \frac{1}{3 \cdot x - 6 \cdot y(x)^2 - y(x) + 5};$$

$$ode1 := \frac{d}{dx} y(x) = \frac{1}{3x - 6y(x)^2 - y(x) + 5}$$

>

>

> $dsolve(\{ode1, y(0) = 1\}, y(x));$

$$y(x) = RootOf\left(18\,\_Z^2\,e^3 - 9\,e^3 x + 15\,\_Z e^3 - 23\,e^{3\,\_Z} - 10\,e^3\right)$$

## Maple Example 9.16

**restart**

> $with(Student[NumericalAnalysis])\ :$

> 

$$f := (x, y) \rightarrow \frac{1}{\left(3 \cdot x - 6 \cdot y^2 - y + 5\right)};$$

$$f := (x, y) \rightarrow \frac{1}{3x - 6y^2 - y + 5}$$

> 

$$x[0] := 0; x[-1] := 0;$$

$$x_0 := 0 \qquad x_{-1} := 0$$

$x_0 := 0$   > $y[0] := 1; y[-1] := 0;$

$$y_0 := 1$$

$y_{-1} := 0$   $h := 0.1;$

$h := 0.1;$   $h := 0.1$

$h := 0.1;$   > $NN := 10;$
$NN := 10$   > $Digits := 8;$

$NN := 10$   $Digits := 8$

> 

**for** $n$ **from** 1 **to** $NN$ **do** $x[n] := x[n-1] + h; k1 := f(x[n-1], y[n-1]); k2 := f\left(x[n-1]\right.$

$+ \left(\frac{1}{2}\right) \cdot h, y[n-1] + \left(\frac{1}{2}\right) \cdot h \cdot k1\right); k3 := f\left(x[n-1] + \left(\frac{1}{2}\right) \cdot h, y[n-1] + \left(\frac{1}{2}\right) \cdot h\right.$

$\left. \cdot k2\right); k4 := f(x[n-1] + h, y[n-1] + h \cdot k3); y[n] := y[n-1] + \left(\frac{1}{6}\right) \cdot h \cdot (k1 + k2$

$+ k3 + k4);$ **end do;**

$x_1 := 0.1, k1 := -\frac{1}{2}, k2 := -0.65412919, k3 := -0.69869767, k4 := -1.2180511,$

$y_1 := 0.94881870 \cdot x_2 := 0.2, k1 := -0.95205426, k2 := -3.0830170, k3 := 1.1542676,$

$$k4 := -0.44248639, y_2 := 0.89343053.$$

$x_3 := 0.3, k1 := -12.086170, k2 := 0.20164019, k3 := -19.409184, k4 := 2.7464794,$

$y_3 := 0.41764328 \cdot x_4 := 0.4, k1 := 0.22543842, k2 := 0.22137238, k3 := 0.22131117,$

$$k4 := 0.21739998, y_4 := 0.43240198.$$

$x_5 := 0.5$ ,$k1 := 0.21524961$ ,$k2 := 0.21148503$ ,$k3 := 0.21143186$ ,$k4 := 0.20780363$ ,
$$y_5 := 0.44650148$$

$x_6 := 0.6$ ,$k1 := 0.20587497$ ,$k2 := 0.20237862$ ,$k3 := 0.20233223$ ,$k4 := 0.19895677$ ,
$$y_6 := 0.45999386$$

$x_7 := 0.7$ ,$k1 := 0.19722154$ ,$k2 := 0.19396523$ ,$k3 := 0.19392458$ ,$k4 := 0.19077612$ ,
$$y_7 := 0.47292532$$

$x_8 := 0.8$ ,$k1 := 0.18921030$ ,$k2 := 0.18616990$ ,$k3 := 0.18613415$ ,$k4 := 0.18319047$ ,
$$y_8 := 0.48533707$$

$x_9 := 0.9$ ,$k1 := 0.18177355$ ,$k2 := 0.17892822$ ,$k3 := 0.17889665$ ,$k4 := 0.17613850$ ,
$$y_9 := 0.49726602$$

$x_{10} := 1.0$ ,$k1 := 0.17485290$ ,$k2 := 0.17218453$ ,$k3 := 0.17215656$ ,$k4 := 0.16956711$ ,
$$y_{10} := 0.50874537$$

In Figs. (**9.94** to **9.28**), we investigate the behavior of solutions to some IVPs using sliders. The slider help us to see how the solution to a problem is sensitive to a parameter change.

> *with*(*Student*[*NumericalAnalysis*]) :

>

$$Explore\left( RungeKutta\left( diff(y(x), x) = \frac{1}{(3 \cdot x - 6 \cdot y(x)^2 - y(x) + 5)}, y(0) = 1, x = a, output \right.\right.$$
$$\left.\left. = plot \right)\right)$$

## Maple Example 9.17

> *restart*

> *with*(*Student*[*NumericalAnalysis*]) :

>

$$Explore\left( RungeKutta\left( diff(y(x), x) = \frac{1}{(3 \cdot x - 6 \cdot y(x)^2 - y(x) + 5)}, y(b) = 1, x = 1, output \right. \right.$$
$$\left. \left. = plot \right) \right)$$

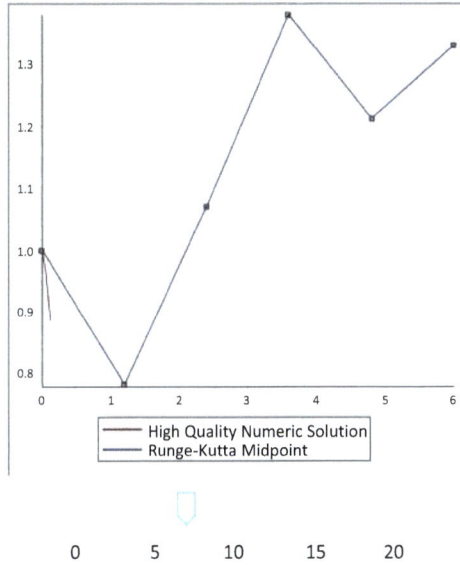

**Fig. (9.24).** Investigating the behaviour of solution of the IVP using slider.

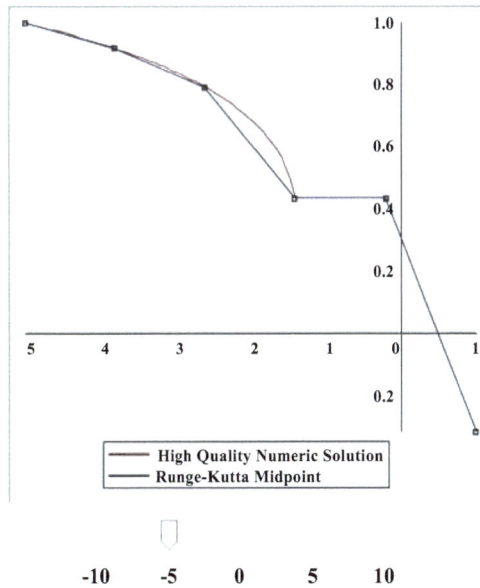

**Fig. (9.25).** Investigating the behaviour of solution of a problem using slider.

> *Explore*(*RungeKutta*(*del, ic, t = a, output = plot*))

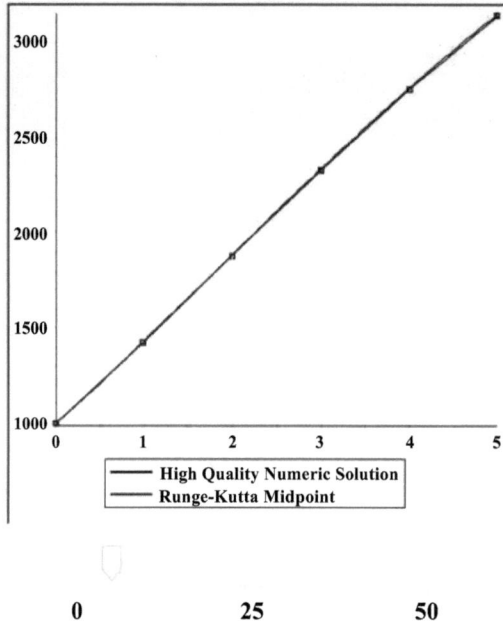

> *Explore*(*RungeKutta*(*del, ic, t = a, output = plot*))

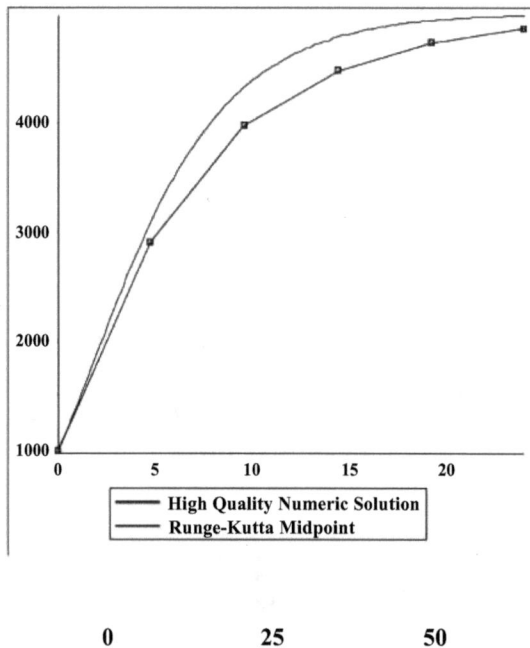

**Fig. (9.26).** Investigating the behavior of solution of a problem using slider.

> $Explore(RungeKutta(de1, ic, t = a, output = plot))$

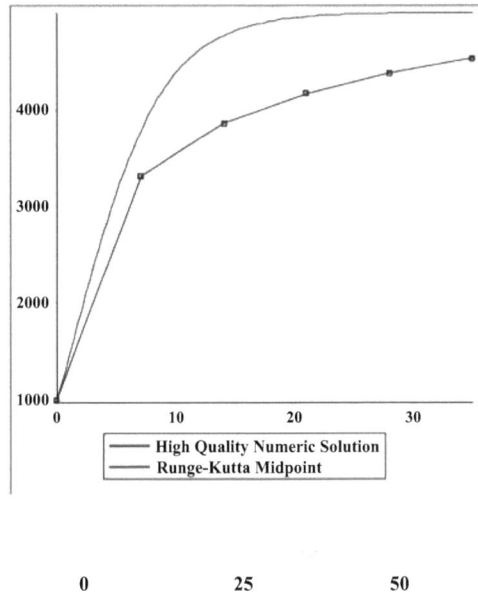

**Fig. (9.27).** Investigating the behavior of solution of a problem using slider.

> $Explore(RungeKutta(de1, ic, t = a, output = plot))$

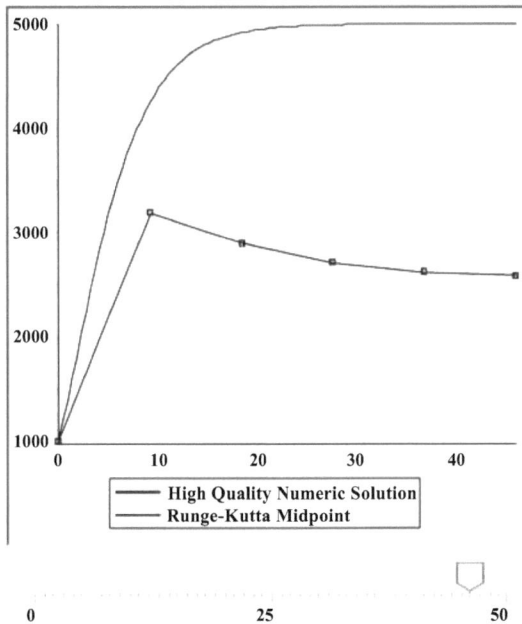

**Fig. (9.28).** Investigating the behavior of solution of a problem using slider.

## Maple Example 9.18

$>c := -1.5; d := 1.23; A := 1.33;$

$c := -1.5$

$d := 1.23$

$A := 1.33$

$>$
$$de2 := diff(V(t), t) = c \cdot V(t) + d \cdot V(t) \cdot \exp(-A \cdot V(t));$$

$$de2 := \frac{d}{dt} V(t) = -1.5\, V(t) + 1.23\, V(t)\, e^{-1.33\, V(t)}$$

$> ic := V(0) = 100;$

$ic := V(0) = 100$

$> dsolve(\{de2, ic\}, V(t))$

$$V(t) = RootOf\left(3t - 3\left(\int_{\_b}^{\_Z} \frac{100}{3\_a\left(-50 + 41\, e^{-\frac{133}{100}\_a}\right)} d\_a\right) + 100\left(\int_{\_b}^{100} \frac{1}{\_a\left(-50 + 41\, e^{-\frac{133}{100}\_a}\right)} d\_a\right)\right)$$

$> with(Student[NumericalAnalysis]):$

$> RungeKutta(de2, ic, t = 200, output = plot)$

## Maple Example 9.19

$>$ Digits: $= 16$:

$>$ dsn16:= dsolve({diff(y(t),t,t)+y(t),y(0)=0,D(y)(0)=1}, numeric, abserr=1e-12, relerr=1e-12, maxfun=0);

$dsn16 := \mathbf{proc}(x\_rkf45) \ ... \ \mathbf{end\ proc}$

\> tt := time():

\> dsn16 (50*Pi);

$$\left[ t = 157.0796326794896, y(t) = -4.7960784\ 10^{-13}, \frac{\mathrm{d}}{\mathrm{d}t}\, y(t) = 1.000000000046069 \right]$$

\> *t_sw := time( ) - tt;*

*t_sw := 19.797*

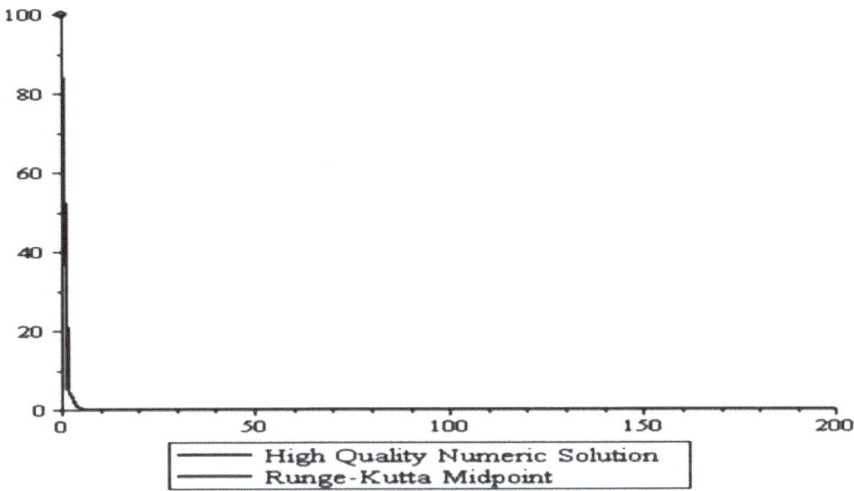

**Fig. (9.29).** Numerical solution of a problem comparing with Runge-Kutta midpoint method.

\> # Run with hardware precision

\> Digits := 10:

\> dsn10 := dsolve({diff(y(t),t,t)+y(t),y(0)=0,D(y)(0)=1}, numeric, abserr=1e-12, relerr=1e-12, maxfun=0);

*dsn10 := **proc**(x_rkf45) ... **end proc***

\> tt := time():

\> dsn10(50*Pi);

$$\left[ t = 157.079632679490, y(t) = -3.55645553154829\ 10^{-14}, \frac{\mathrm{d}}{\mathrm{d}t}\, y(t) = 1.00000000004606 \right]$$

> t_hw := time()-tt;

*t_hw* := 0.125

#Run with hardware precision and compile

> dsn10 := dsolve({diff(y(t),t,t)+y(t),y(0)=0,D(y)(0)=1}, numeric, abserr=1e-12, relerr=1e-12, maxfun=0, compile=true);

*dsn10* := **proc**(*x_rkf45*) ... **end proc**

> tt := time():

> dsn10(50*Pi);

$$\left[ t = 157.079632679490, y(t) = -3.55645553154829 \ 10^{-14}, \frac{d}{dt} y(t) = 1.00000000004606 \right]$$

> t_hc:= time ()-tt;

*t_hc* := 0.031

#Compare:

> t_sw/t_hc, t_hw/t_hc;

638.6129032, 4.032258065

## Maple Example 9.20

> *restart*

> *with(plots)* :

> *with(Student[NumericalAnalysis])* :

> *A := Array(1..2)* :

> *A[1] := Euler(diff(z(t), t) = 1/(2\*t−3\*z(t) + 5), z(0) = 1, t = 1, output = plot);*

*A₁* := *PLOT(...)*

> $A[2] := RungeKutta(diff(z(t), t) = 1/(2*t-3*z(t)+5), z(0) = 1, t = 1, output = plot);$

$A_2 := PLOT(...)$

> $display(A)$

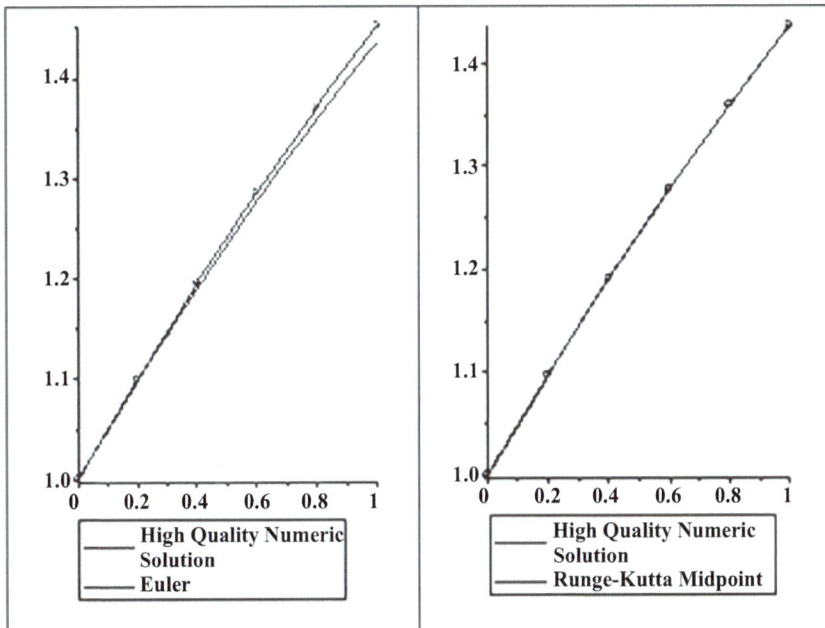

## CONCLUSION

There are several emerging complex nonlinear ordinary differential equations (ODEs) arising from models from science, business and technology, which could only be solved by the use of computers. The analytic or exact solutions to most nonlinear ordinary differential equations cannot be easily found even with the applications of symbolic programming. The needs for computer-based solutions to those ODEs problems are becoming increasingly important. Applications of numerical methods and development of numerical simulations are everywhere present in most researches in engineering, economics and life science these days.

The chapter gives the basic numerical treatment on numerical methods for obtaining solutions to ODEs to complement analytical solutions in the other chapters. Methods discussed are Taylor series, Euler, Modified Euler's, Runge-Kutta, Adams-Bash forth-Moulton and Milne numerical methods together with some Maple examples. Numerical simulation procedures using Maple software are presented and applied to the following problems: HIV/AIDS, Fitzhugh, and Fitzhugh-Nagumo, sickle cell anemia, zooplankton-fish, Gompertz tumor and neural firing models.

## REFERENCES

[1]   S.O. Ale, and B.O. Oyelami, "Impulsive systems and their applications", *Int. J. Math. Educ. Sci. Technol.,* vol. 31, no. 4, pp. 539-544, 2000.
      http://dx.doi.org/10.1080/002073900412642

[2]   A.M. Edwards, ""Adding detritus to a nutrient-phytoplankton-zooplankton model". Dynamical systems approach", *J. Plankton Res.,* vol. 23, no. 4, pp. 389-413, 2001.
      http://dx.doi.org/10.1093/plankt/23.4.389

[3]   H. Malchow, B. Radtke, M. Kallache, A.B. Medvinsky, D.A. Tikhonov, and S.V. Petrovskii, "Spatio-temporal pattern formation in coupled models of plankton dynamics and fish school motion", *Nonlinear Anal. Real World Appl.,* vol. 1, no. 1, pp. 53-67, 2000.
      http://dx.doi.org/10.1016/S0362-546X(99)00393-4

[4]   H.K. Dass, and Er. Rajnish Verma, *Higher Engineering Mathematics.* S. Chand & Company Ltd, Ram Nagar: New Delhi, 2006.

[5]   *Maple 18 Software,* Maplesoft, a division of Waterloo Maple Inc: Canada, 2013.www.maplesoft.com

[6]   B.O. Oyelami, "Comments on sickle cell anemia through variation and oscillatory methods", *Caribb. J. Sci. Technol.,* vol. 1, pp. 43-51, 2013.

[7]   B.O. Oyelami, and S.O. Ale, "Impulsive model for the absorption of oxygen by the red blood cells in the presence of nitric oxide yielding drugs", *African J. Phys.,* vol. 3, 2010.

[8]   B.O. Oyelami, "Oxygen and hemoglobin pair model for sickle cell anemia patients,", *African J. Phys.,* vol. 2, pp. 132-143, 2009.

[9]   B.O. Oyelami, N.H. Manjak, J.A. Ogidi, and A.M. Mustapha, "Stability property of some neural firing model using Lyapunov function", *JOLORN.,* vol. 4, no. 1, pp. 44-58, 2003.

# APPENDIX A

# Brief Highlights about Maple Software

**Abstract:** Many real-life problems can be solved through modeling and simulation and Maple 2022 is the world-leading software used by mathematicians, physicists, economists, engineers, and educators for the problem solving task. The power of Maple and the MapleSim software are exploited in this section. We present the starting process with the software and demonstrate the application of the software *via* some selected problems.

**Keywords:** 2D, 3D plots, Animation, C Codes, Hybrid computations, Maple, MapleSim, MapleSim, Monte Carlo Simulation, Numerical, Symbolic, Simulation.

## A.1. POWER OF MAPLE

Maple has the most powerful Math engine, and smart document interface, along with Maple add-in and grid computing facilities for symbolic, numerical, and hybrid computation, sophisticated 2D, 3D plotting and animation, and document and word processing tools.

Furthermore, Maple T.A (Test and Assessment) has an E-learning solution. Maple T A is an easy-to-use web-based system for creating tests and assignments and automatically assessing students' responses and performance. It has Maple T.A. placement Test suite to deliver tests online which reduces the cost of administration and marking examinations using paper type.

### A.1.1 Maple Net

- Maple Net: A facility that allows easy sharing of Maple documents, calculator and technical application. There is also MapleSim 2022 for the simulation of engineering and real-life processes.

### A.1.2 Calculus Kits

- Calculus Kits are for students and teachers to interact with each other while solving mathematical problems.

### A.1.3 Users

- Maple is software that can be used by mathematicians, physicists, engineers, chemists, social scientists and educators.

**Benjamin Oyediran Oyelami**
**All rights reserved-© 2024 Bentham Science Publishers**

## 1.4 Maple Portal

- Maple makes use of what is called Portals. The Maple Portal is designed as a starting place for any Maple user. There are 3 types of portals in Maple that are related to our study in this textbook and these are:
- Portal for Engineers: which contains tools used by Engineers in solving mathematical problems. Engineering packages contain a dynamical system toolbox, scientific constants, scientific error analysis, tolerance and units.
- Portal for Students: Student packages are available for the following topics: Pre-calculus, calculus, vector calculus, differential equations, linear Algebra, and multivariate calculus.
- Portal for Math Educators: This portal contains information and tools for education, assessment, Maple Test and assessment of students. This portal contains student packages that allow instructors to deliver the course contents effectively; give students insight into understanding basic mathematical concepts and enhance their problem-solving practical skills. There is also a survival kit to enhance students' mathematical mastery of topics in the portal for students

### A.1.5 Help Resources and Maple Tour

Maple also has Help Resources and Maple Tour to give tutorials on how to use the resources in Maple and Help system to help the users out of perceived problems and many examples on how to use maple resources. There is also a Quick reference card. This card gives vital information on how to make use of resources like the type of modes for creating documents in Maple. It also gives information on Toggle Math/Text entry mode, how to evaluate math expressions and display results in line; common operations available in the Maple in both document and worksheet Modes; 2-D math editing operations, keyboard shortcuts, and operations plotting and animation.

### A.1.6 User Manuals and Web links

User Manuals: This manual gives comprehensive information about Maple, tutorials, and examples on Maple. The manual contains how to get started with maple toolboxes, the user manual and the programming guild.

Web links: This is the hyperlink to Maple soft Company, which is the developer and marketer of Maple software. The links provide information and registration of the company, and show how to register and take part on webinars, an online seminar series. It also provides information on how to get online resources on Maple.

## A.2 Getting Started

### A.2.1 Maple Tutorial

Maple tutorial helps to get started with the software, learn about the tools available in Maple, and lead you through a series of problems. It guides you on how to enter simple expression, functions, matrices, complex numbers, and evaluate expression and plotting functions.

Maple has so many interesting modelling and simulation facilities as we are not going to make a discussion on them but we will demonstrate their applications in Maple Examples.

### Examples on Graphs and Animations

### Example A1

To plot the graph of sine function in the worksheet mode, type in the command:

**> Plot (sin (2\*x),x =-Pi..Pi, thickness=2);**

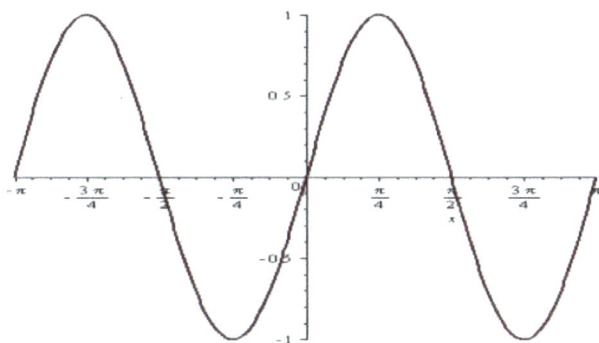

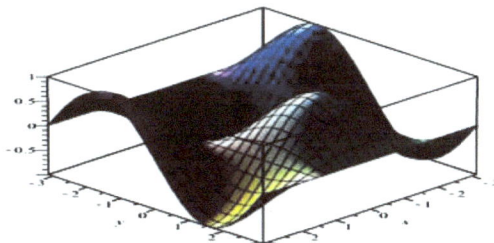

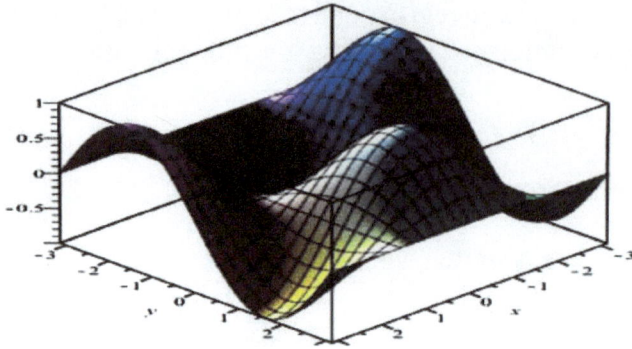

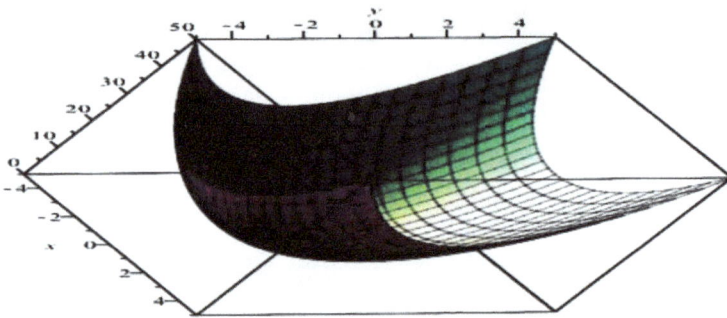

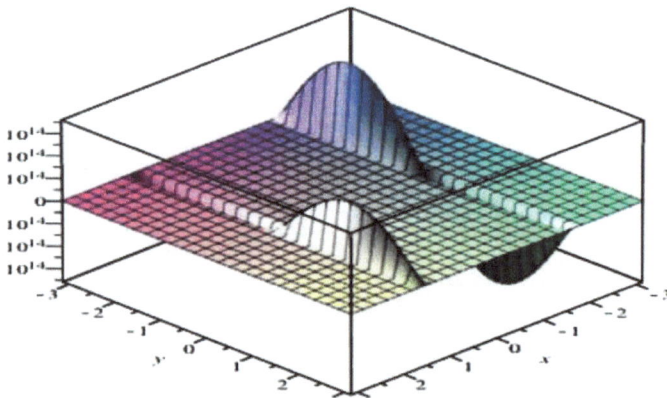

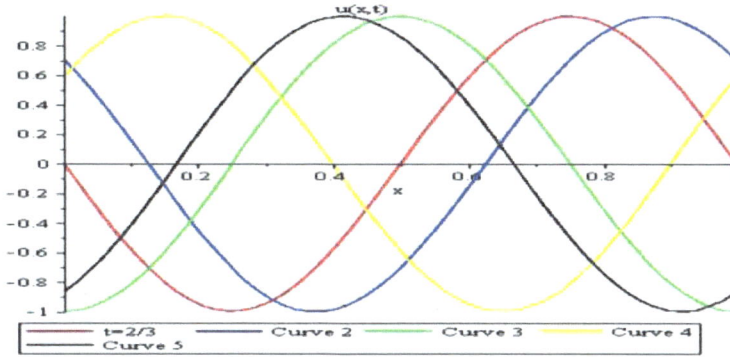

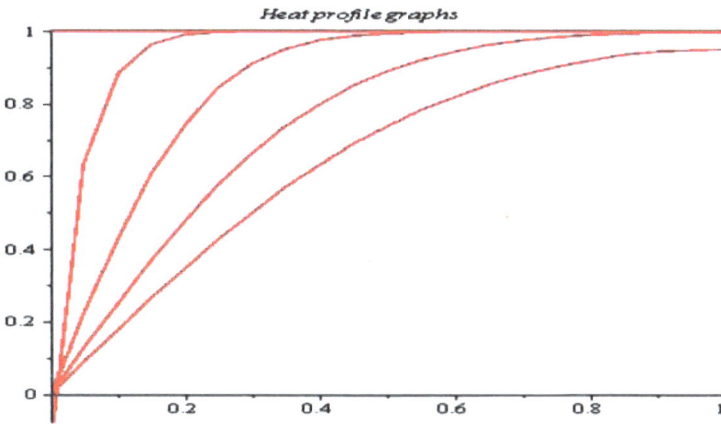

Heat profile graphs

$plot(\sin(2 \cdot x), x = -\text{Pi} .. \text{Pi}, thickness = 30);$

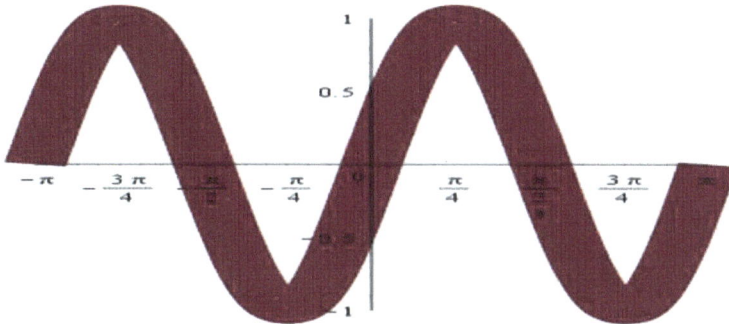

$> f := 2 \cdot \sin(2 \cdot \text{Pi} \cdot x) + 2 \cdot x;$

$f := 2 \sin(2 \pi x) + 2 x$

$> plot(f, x = -10 .. 10, thickness = 10);$

$>animate(plot, [\sin(t), t =-10..x], x = 0..Pi);$

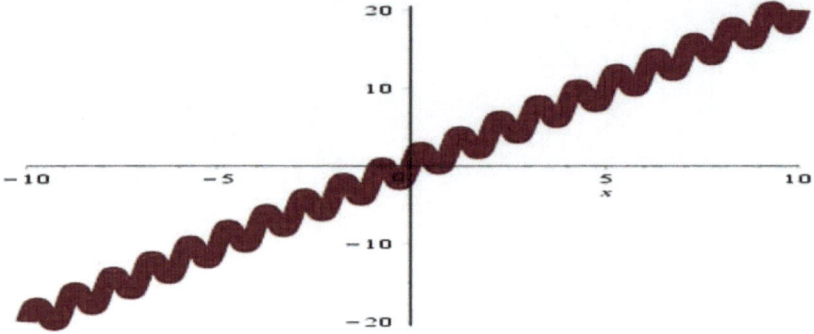

$>$

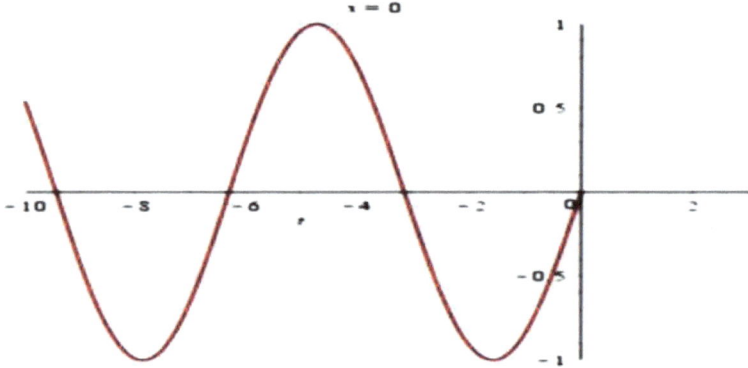

$> animate(plot, [2\cdot\sin(2\cdot Pi\cdot t) + 2\cdot t, t =-10..x], x = 0..Pi);$

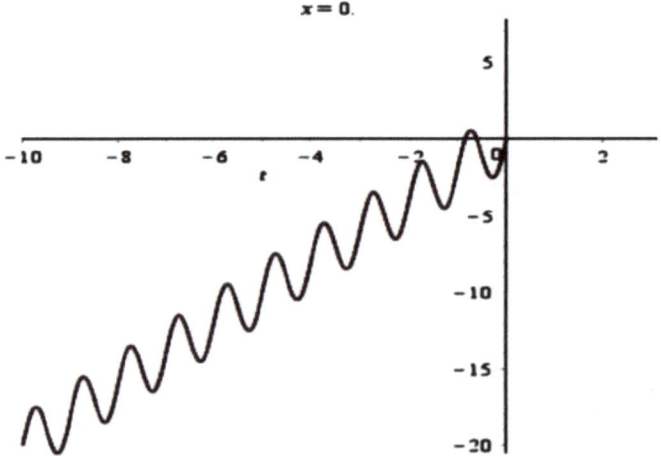

\> *restart*

\> *with*(*plots*) :

\> *animate3d*(cos(*t·x*)·sin(3·*t·y*), *x* =-Pi ..Pi, *y* =-Pi ..Pi, *t* = 1 ..2);

We can extend the plot to 3D using document mode: type in the following two dimension function w=w(x, y) and highlight the equation, right click the 3D plot, we have:

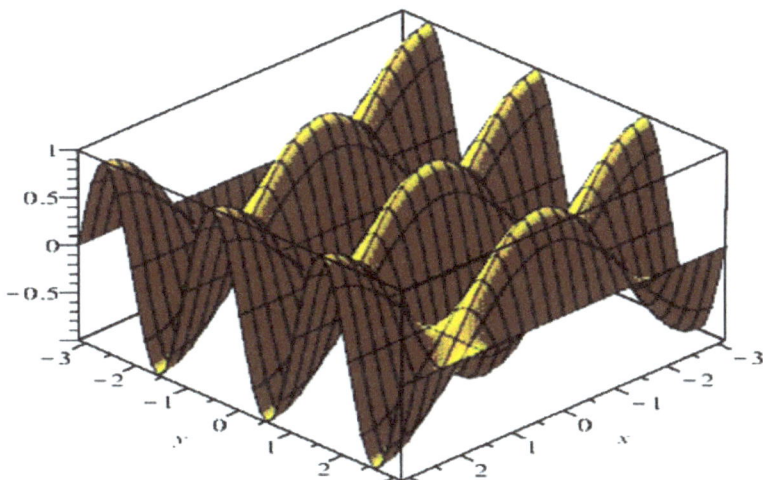

\>

\> *restart*

\> *with*(*plots*) :

\> *with*(*DEtools*) :

In the document mode, type the equation and highlight it and right click to select the 2D plot, then we have the plot:

$y = 2x^2 + 3x + 9$  →

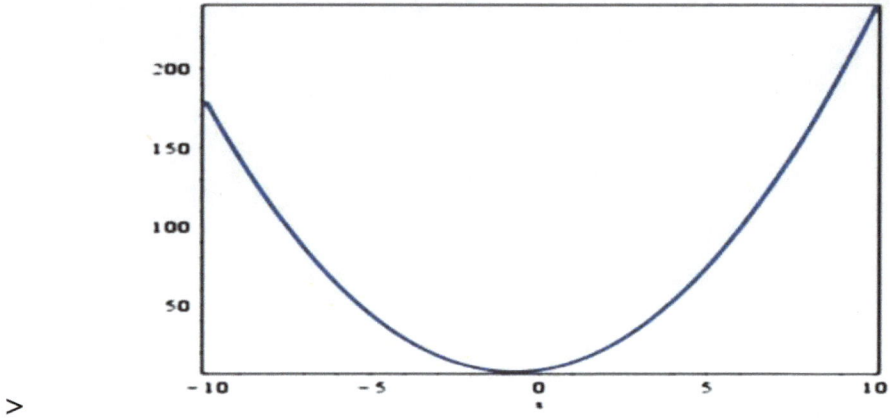

>

>We can also replicate the above plot using worksheet mode by typing in the equation and right- click and select 3D plot. You can also use the plot builder to have a variety of 3D-plots and even animate the plots too.

$$w = x^2 + y^2 + 3xy + \frac{2xy}{x^2 + \sin(xy)} \rightarrow$$

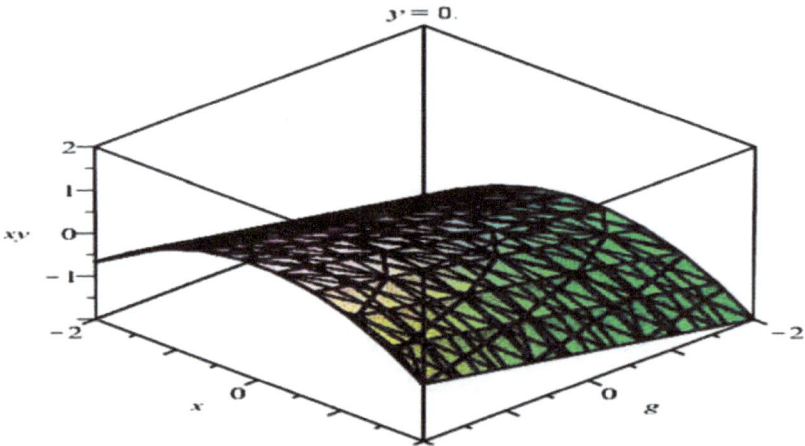

>**w=3x^2** $+ 3xy + y^2 + \dfrac{2xy}{x^2 + \sin(xy)}$;

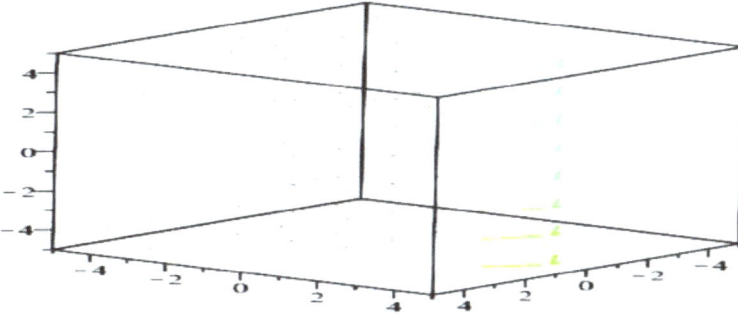

>

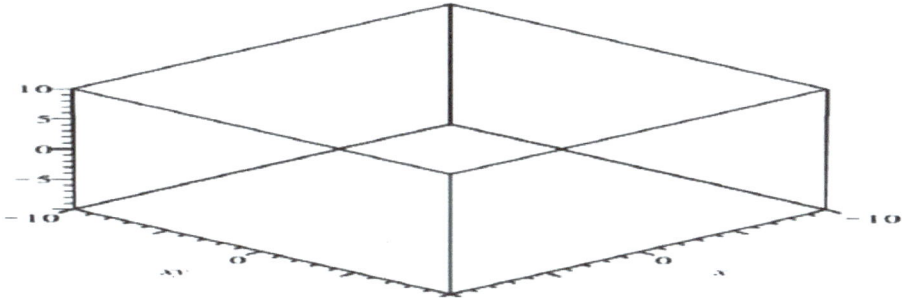

>

**Using worksheet mode, you can animate a plot using the command with (plots) together with animate3d. For example, type in:**

>*with*( *plots* ) :

*animate3d*( $\sin(t \cdot x \cdot y) \cdot \cos(t \cdot x \cdot y)$, $x = -\text{Pi}..\text{Pi}$, $y = -\text{Pi}..\text{Pi}$, $t = 0..1$ )

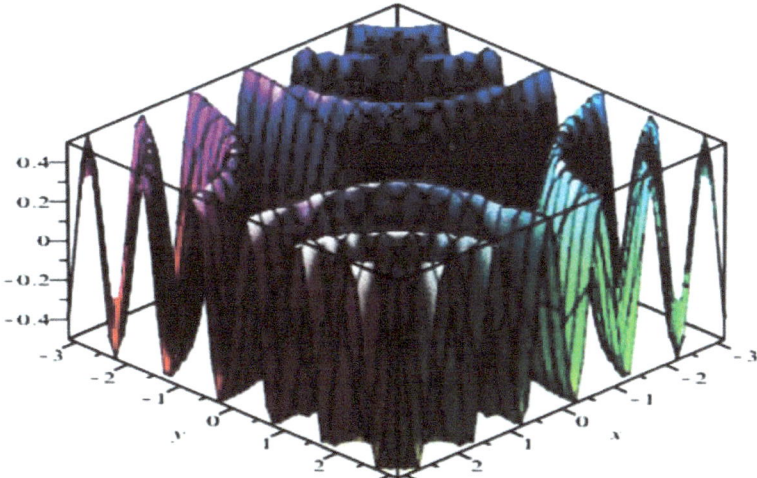

Highlight the 3D-plot and select the type of animation, whether short-time animation or continuous one. In Maple software, the memory can be cleared using 'restart'.

*restart*

*with( plots )* :

*animate3d(* $\exp(t \cdot x \cdot y) \cdot \sin(t \cdot x \cdot y) \cdot \cos(t \cdot x \cdot y), x = -\text{Pi}..\text{Pi}, y = -\text{Pi}..\text{Pi}, t = 0..1$ )

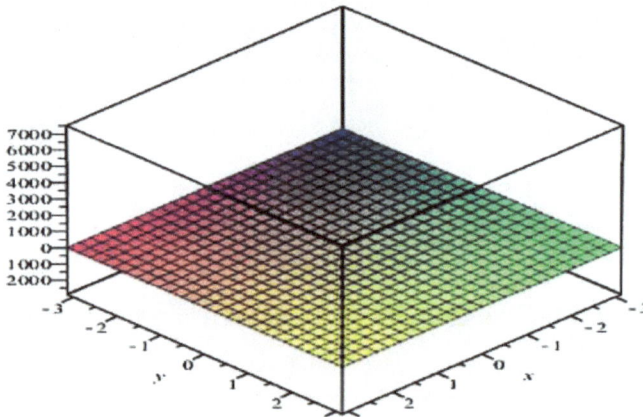

Maple contains several facilities for computation using Linear Algebra. Type in with (LineraAlgebra) with 'semicolon' to display the linear algebra facilities in the maple software. We can suppress this by using colon as usual.

> *with( LinearAlgebra )* ;

**In the worksheet mode, a vector and a matrix can be typed in as follows:**

**x=Vector ([1, 0,-2, 3);**

$$
x := \begin{bmatrix} 1 \\ 0 \\ -2 \\ 3 \end{bmatrix}
$$

[ *&x, Add, Adjoint, BackwardSubstitute, BandMatrix,*
*Basis, BezoutMatrix, BidiagonalForm, BilinearForm,*
*CARE, CharacteristicMatrix,*
*CharacteristicPolynomial, Column,*
*ColumnDimension, ColumnOperation, ColumnSpace,*
*CompanionMatrix, CompressedSparseForm,*
*ConditionNumber, ConstantMatrix, ConstantVector,*
*Copy, CreatePermutation, CrossProduct, DARE,*
*DeleteColumn, DeleteRow, Determinant, Diagonal,*
*DiagonalMatrix, Dimension, Dimensions,*
*DotProduct, EigenConditionNumbers, Eigenvalues,*
*Eigenvectors, Equal, ForwardSubstitute,*
*FrobeniusForm, FromCompressedSparseForm,*
*FromSplitForm, GaussianElimination,*
*GenerateEquations, GenerateMatrix, Generic,*
*GetResultDataType, GetResultShape,*
*GivensRotationMatrix, GramSchmidt, HankelMatrix,*
*HermiteForm, HermitianTranspose, HessenbergForm,*
*HilbertMatrix, HouseholderMatrix, IdentityMatrix,*
*IntersectionBasis, IsDefinite, IsOrthogonal, IsSimilar,*
*IsUnitary, JordanBlockMatrix, JordanForm,*
*KroneckerProduct, LA_Main, LUDecomposition,*
*LeastSquares, LinearSolve, LyapunovSolve, Map,*
*Map2, MatrixAdd, MatrixExponential,*
*MatrixFunction, MatrixInverse,*
*MatrixMatrixMultiply, MatrixNorm, MatrixPower,*
*MatrixScalarMultiply, MatrixVectorMultiply,*
*MinimalPolynomial, Minor, Modular, Multiply,*
*NoUserValue, Norm, Normalize, NullSpace,*
*OuterProductMatrix, Permanent, Pivot, PopovForm,*
*ProjectionMatrix, QRDecomposition, RandomMatrix,*
*RandomVector, Rank, RationalCanonicalForm,*
*ReducedRowEchelonForm, Row, RowDimension,*
*RowOperation, RowSpace, ScalarMatrix,*
*ScalarMultiply, ScalarVector, SchurForm,*
*SingularValues, SmithForm, SplitForm,*
*StronglyConnectedBlocks, SubMatrix, SubVector,*
*SumBasis, SylvesterMatrix, SylvesterSolve,*
*ToeplitzMatrix, Trace, Transpose, TridiagonalForm,*
*UnitVector, VandermondeMatrix, VectorAdd,*
*VectorAngle, VectorMatrixMultiply, VectorNorm,*
*VectorScalarMultiply, ZeroMatrix, ZeroVector, Zip* ]

>A:=Matrix([[1, 2, 0, 3], [0, 0, -1, 4], [0, 0, -3, 2], [2, 1, 0, 2]]);

$$A := \begin{bmatrix} 1 & 2 & 0 & 3 \\ 0 & 0 & -1 & 4 \\ 0 & 0 & -3 & 2 \\ 2 & 1 & 0 & 2 \end{bmatrix}$$

The element on the first row fourth colon can be displayed by typing in:

> $A[1, 4]$;

3

> $A[3, 3]$;

-3

A matrix A can be multiplied by itself using the code:

> $A.A$;

$$\begin{bmatrix} 7 & 5 & -2 & 17 \\ 8 & 4 & 3 & 6 \\ 4 & 2 & 9 & -2 \\ 6 & 6 & -1 & 14 \end{bmatrix}$$

Matrix A can be post multiplied using the vector x as follows:

> $A.x$

$$\begin{bmatrix} 10 \\ 14 \\ 12 \\ 8 \end{bmatrix}$$

B: =Matrix ([[1,2],[5,7],[3,5], [0,3]]);

$$B := \begin{bmatrix} 1 & 2 \\ 5 & 7 \\ 3 & 5 \\ 0 & 3 \end{bmatrix}$$

> *A.B;*

$$\begin{bmatrix} 11 & 25 \\ -3 & 7 \\ -9 & -9 \\ 7 & 17 \end{bmatrix}$$

> *B.A;*

Error, (in LinearAlgebra:-Multiply) first matrix column dimension (2) <> second matrix row dimension (4)

> $h := i \rightarrow i^2;$

$$h := i \rightarrow i^2$$

> **y** := **Vector(8, h);**

$$y := \begin{bmatrix} 1 \\ 4 \\ 9 \\ 16 \\ 25 \\ 36 \\ 49 \\ 64 \end{bmatrix}$$

> *# generate the HilbertMatrix*

> **H:= Matrix(5,5, (i,j) -> 1/(i+j-1));**

$$H := \begin{bmatrix} 1 & \dfrac{1}{2} & \dfrac{1}{3} & \dfrac{1}{4} & \dfrac{1}{5} \\[2ex] \dfrac{1}{2} & \dfrac{1}{3} & \dfrac{1}{4} & \dfrac{1}{5} & \dfrac{1}{6} \\[2ex] \dfrac{1}{3} & \dfrac{1}{4} & \dfrac{1}{5} & \dfrac{1}{6} & \dfrac{1}{7} \\[2ex] \dfrac{1}{4} & \dfrac{1}{5} & \dfrac{1}{6} & \dfrac{1}{7} & \dfrac{1}{8} \\[2ex] \dfrac{1}{5} & \dfrac{1}{6} & \dfrac{1}{7} & \dfrac{1}{8} & \dfrac{1}{9} \end{bmatrix}$$

> $C := \langle\langle 1, 2, 3\rangle | \langle 0, 0, 1\rangle | \langle 0, 0, 1\rangle\rangle;$

$$C := \begin{bmatrix} 1 & 0 & 0 \\ 2 & 0 & 0 \\ 3 & 1 & 1 \end{bmatrix}$$

> # *Find the basis for A and C;*

> *NullSpace*$(A)$;

$\{\ \}$

> *NullSpace*$(C)$;

$$\left\{ \begin{bmatrix} 0 \\ -1 \\ 1 \end{bmatrix} \right\}$$

> # *A has no basis;*

> $d := \langle 1, 2, 0, -1\rangle;$

$$d := \begin{bmatrix} 1 \\ 2 \\ 0 \\ -1 \end{bmatrix}$$

>z:=LinearSolve(A,d);

Warning, inserted missing semicolon at end of statement

$$z := \begin{bmatrix} -\dfrac{6}{5} \\ \dfrac{1}{5} \\ \dfrac{2}{5} \\ \dfrac{3}{5} \end{bmatrix}$$

> E:=IdentityMatrix(4);

Warning, inserted missing semicolon at end of statement

$$E := \begin{bmatrix} 1 & 0 & 0 & 0 \\ 0 & 1 & 0 & 0 \\ 0 & 0 & 1 & 0 \\ 0 & 0 & 0 & 1 \end{bmatrix}$$

> $Determinant(A);$

-30

> $Rank(A);$

4

> *l2ip* := $(f, g) \rightarrow int(f(x) \cdot g(x), x = 0..1)$;

$$l2ip := (f, g) \rightarrow \int_0^1 f(x)\, g(x)\, dx$$

>N:=f ->sqrt(%(f,f));

$$N := f \rightarrow \sqrt{\%(f, f)}$$

> *unassign* ( **'x'** );

> $f := x \rightarrow x \cdot (1 - x)$;

$$f := x \rightarrow x\,(1 - x)$$

>
$$g := x \rightarrow \frac{8}{\pi^3} \cdot \sin(\text{Pi} \cdot x);$$

$$g := x \rightarrow \frac{8 \sin(\pi x)}{\pi^3}$$

> plot({f(x),g(x)},x = 0 .. 1,thickness = 6);

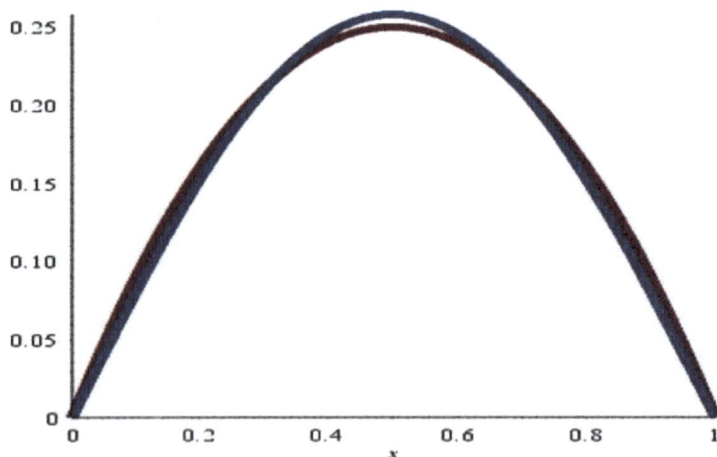

>>> $edn := diff(y(x), x) = -\dfrac{x}{y};$

$edn := \dfrac{d}{dx} y(x) = -\dfrac{x}{y}$

> $with(plots)$

$[$ *animate, animate3d, animatecurve, arrow, changecoords, complexplot, complexplot3d,*
     *conformal, conformal3d, contourplot, contourplot3d, coordplot, coordplot3d, densityplot,*
     *display, dualaxisplot, fieldplot, fieldplot3d, gradplot, gradplot3d, implicitplot,*
     *implicitplot3d, inequal, interactive, interactiveparams, intersectplot, listcontplot,*
     *listcontplot3d, listdensityplot, listplot, listplot3d, loglogplot, logplot, matrixplot, multiple,*
     *odeplot, pareto, plotcompare, pointplot, pointplot3d, polarplot, polygonplot, polygonplot3d,*
     *polyhedra_supported, polyhedraplot, rootlocus, semilogplot, setcolors, setoptions,*
     *setoptions3d, spacecurve, sparsematrixplot, surfdata, textplot, textplot3d, tubeplot* $]$

> $edn := diff(y(x), x) = -\dfrac{x}{y};$

$edn := \dfrac{d}{dx} y(x) = -\dfrac{x}{y}$

> $c := gradplot\left( -\dfrac{x}{y}, x = -10..10, y = -10..10 \right)$

$c := PLOT(...)$

>

> **with(plottools):**

> **with(plots):**

> **c1:= circle([1,1], 1,color=blue):**

> **c2:=circle([1/2,1], 1/2,color=red):**

> **display([c1,c2);**

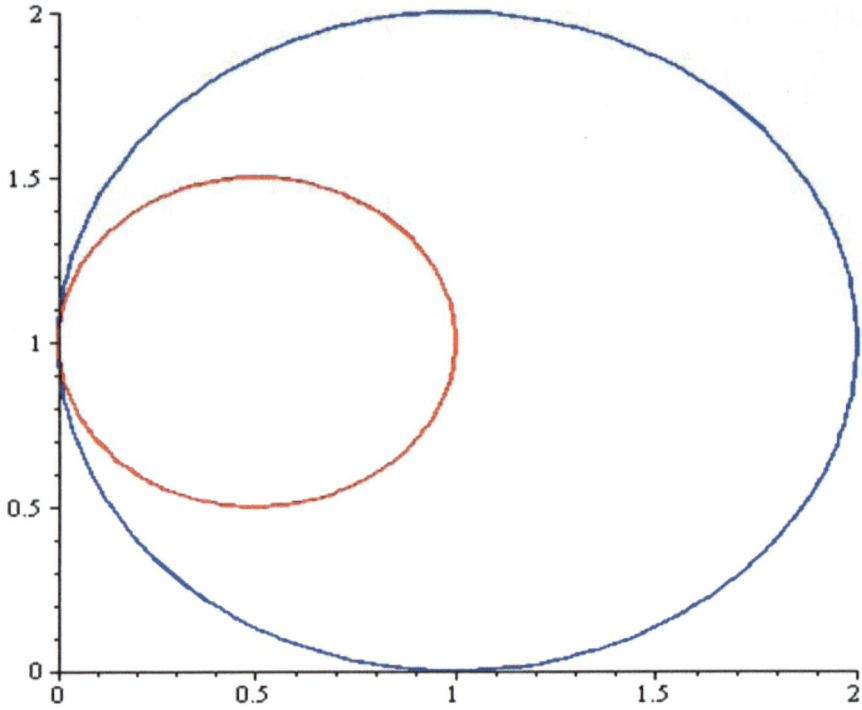

> $display(c, c2, c1); :$

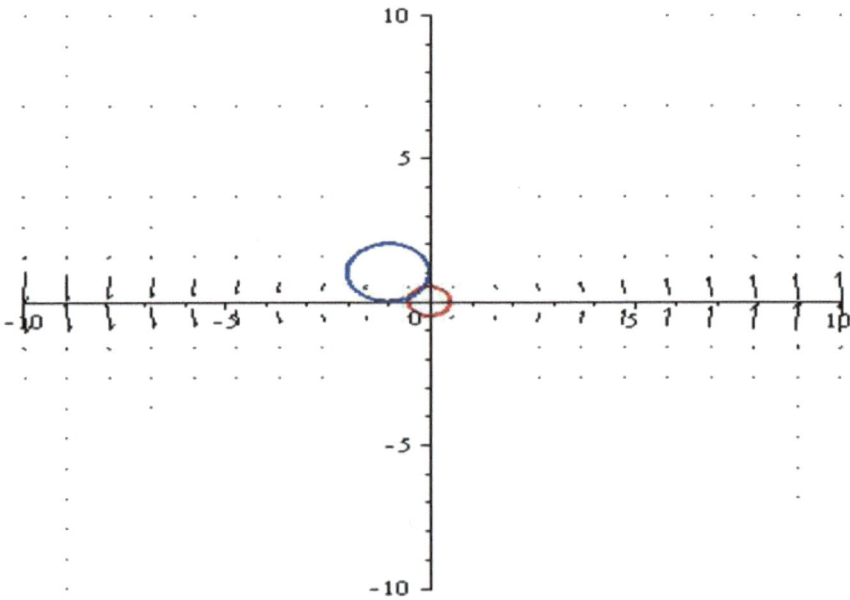

> *with*(*plots*) :

> *c* := *sphere*([1, 1, 1], 3.3) :

> *display*(*c*, *scaling* = *constrained*, *axes* = *boxed*)

> *d* := *sphere*([1, 5, 1], 2) :

> *display*(*c*, *d*, *scaling* = *constrained*, *style* = *patchnogrid*)

> *esphere*([1, 5, 1], 2) :

> *display*(*c*, *e*, *scaling* = *constrained*, *style* = *patchnogrid*)

> 

> :=

> with(**plottools**):

> with(plots):

> c1 := ellipse([1,1], 1, **color=blue**):

> c2 := circle([1/2,1], 1/2, **color=red**):

> display(c1,c2);

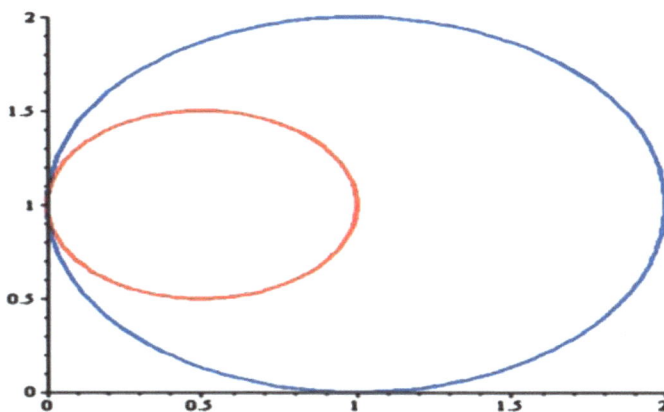

> c3:= ellipse([-1,1], 1,color=blue):

> c4:= circle([-1/2,1], 1/2,color=red):

> with(plottools):

> with(plots):

> display(c1,c2,c3,c4);

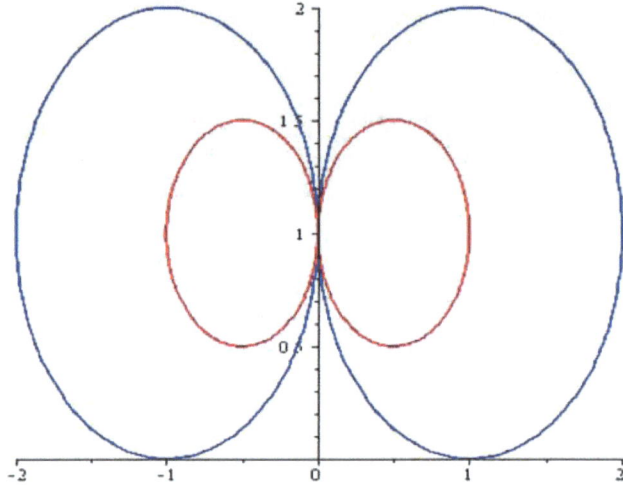

> c2:=circle([1/2,1], 1/2,color=red):

> display(c1,c2);

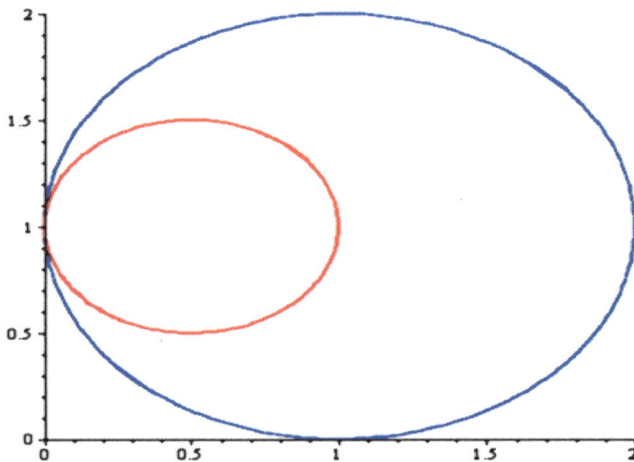

> *restart*

> *with*(*plots*) :

> *dualaxisplot*(*plot*(sin), *plot*(cos))

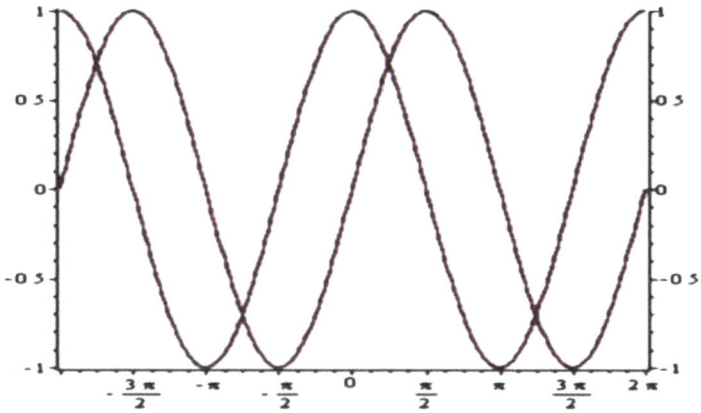

>

$$dualaxisplot\big(inequal(\{x - y \le 5, 0 < x + y\}, x = -10 ..10, y = -10 ..10, optionsexcluded$$
$$= (color = white)), conformal\big(z^2, z = 0 ..5 + 5I\big)\big)$$

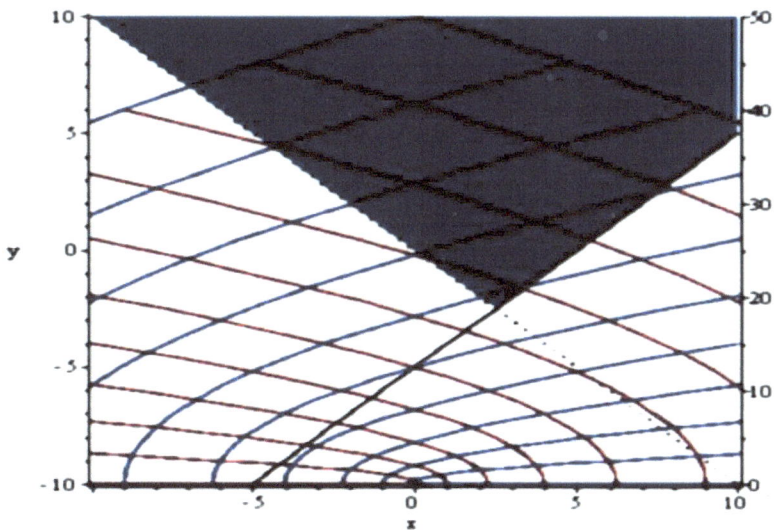

>

$dualaxisplot\left(plot\left(x^2 \cdot \exp(-x), x = 0..10, labels = [x, y], legend = x^2 \cdot \exp(-x)\right), plot\left(x^3, x = 0\right.\right.$
$\left..10, color = blue, labels = \left[x, x^3\right], legend = x^3\right), title = \text{" Plots of two graphs "}\big)$

>

$dualaxisplot\big(animate\big(plot, \left[A\,x^3, color = blue, labels = \left[x, x^3\right]\right], A = 0..1\big), plot\left(x^2, labels = [x,\right.$
$\left.\left. x^2\right]\right)\big)$

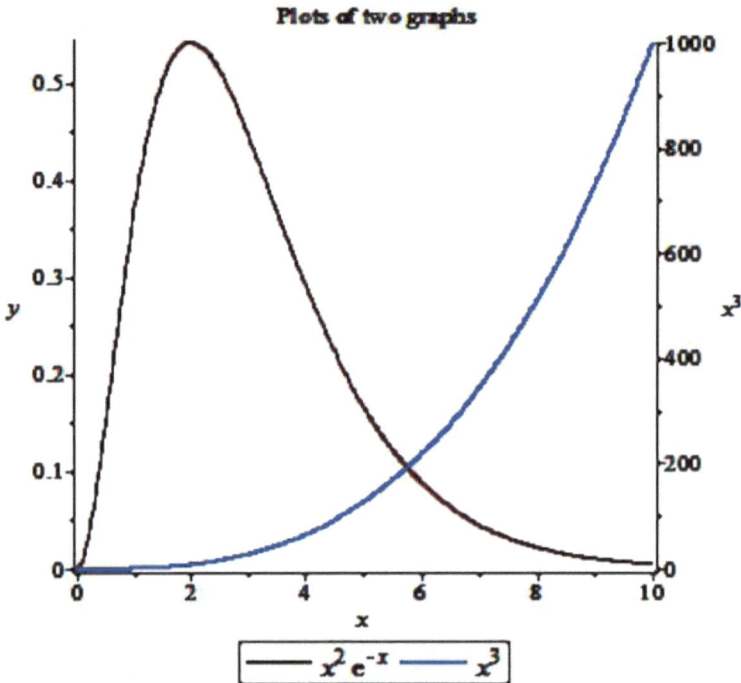

> *restart*

> with(plots,[pareto]):

>pdata:= 'Engine 1'=327,

`Engine 2`= 240,

`Engine 3`=176,

`Wire 1`=105,

**`Wire 2`=43,**

**`Wire 3`=36,**

**Oil=33,**

  **Coils=90,**

**`Gear Box`=61,**

**`Steam line`=50,**

**Others=166]:**

>Fdata:=map(rhs,Pdata):

> Lab:=map(lhs,Pdata):

> > *pareto*(*Fdata, tags = Lab, title = `Plant Problems`*);

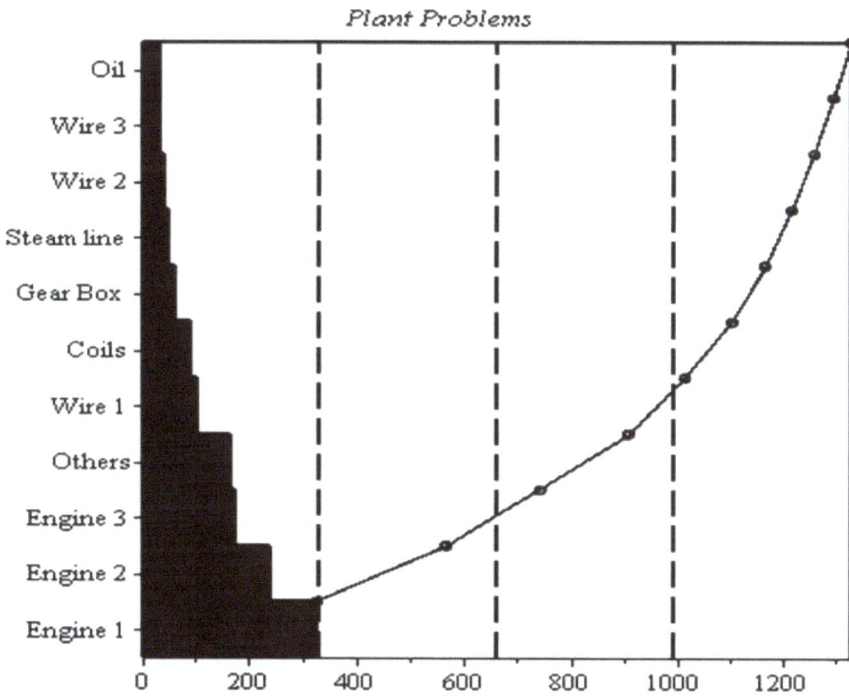

> **Fdata_norm:=map((x,s) -> 100*x/s, Fdata, `+`(op(Fdata))):**

> **pareto(Fdata_norm,   tags=Lab,   misc=Others,   title=`Percentages**  of problems`);

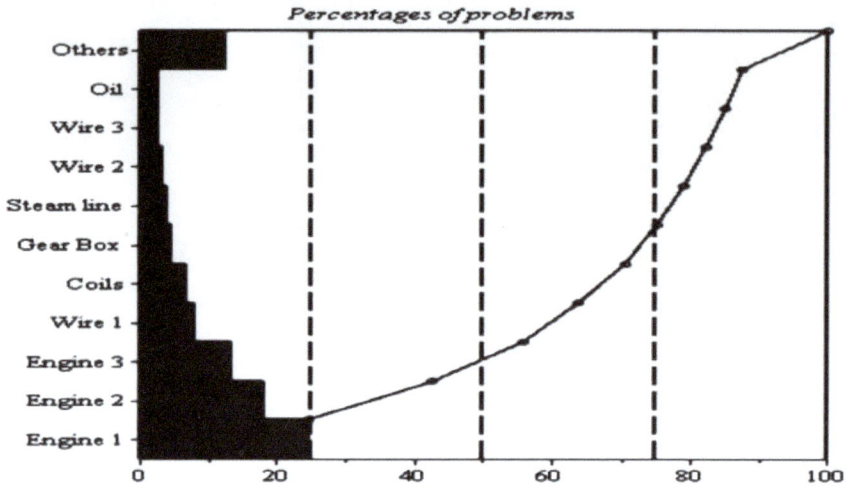

Percentages of problems

> *restart*

> *with*(*plots*) :

> *tubeplot*( [cos(*t*), sin(*t*), 0], *t* = 0 ..2 π, *radius* = 0.5)

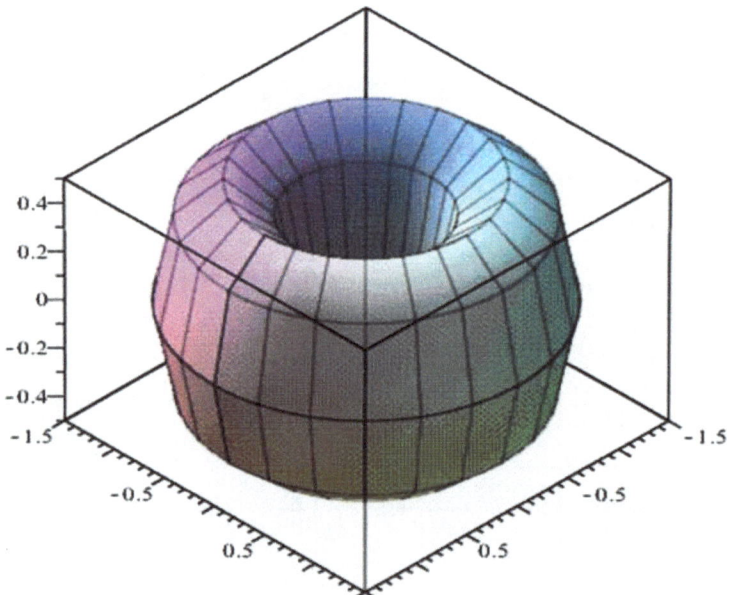

> *tubeplot*( [exp(*t*) · cos(*t*), exp(*t*) · sin(*t*), 0], *t* = 0 ..2 π, *radius* = 0.5)

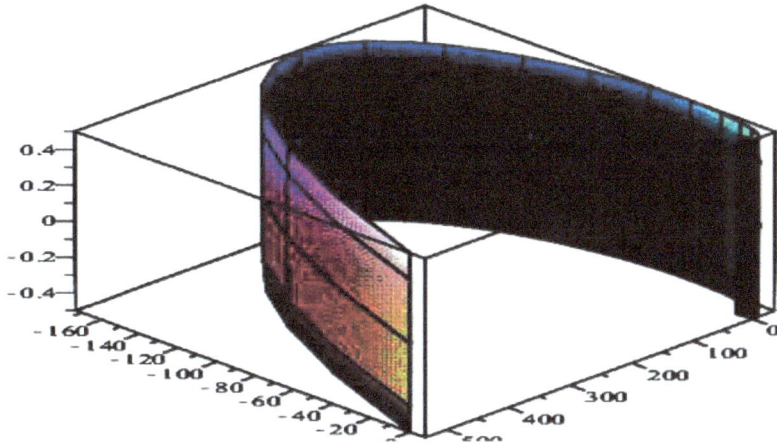

> *tubeplot*( [exp( −*t*) · cos(*t*), exp( −*t*) · sin(*t*), 0], *t* = 0 ..5 π, *radius* = 0.5)

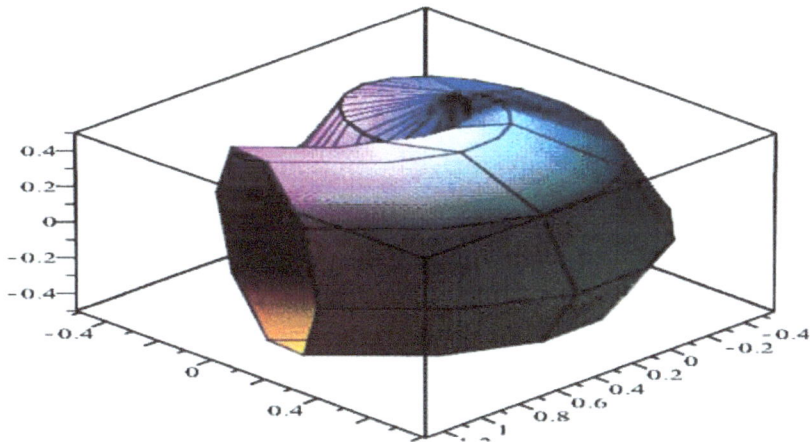

> *restart*

> *with*(*plots*) :

> *polyhedra_supported*( ) :

> *polyhedraplot*( [0, 0, 0], *polytype* = *GyroelongatedPentagonalPyramid*, *scaling* = *constrained*)

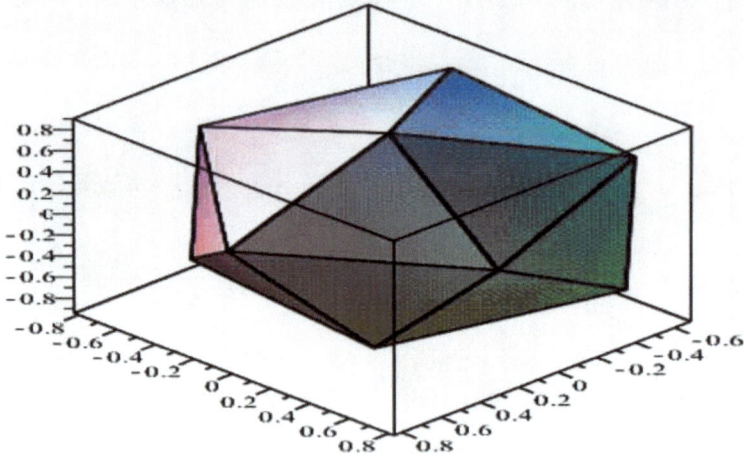

> *polyhedraplot([0, 0, 0], polytype = TriakisIcosahedron, scaling = constrained)*

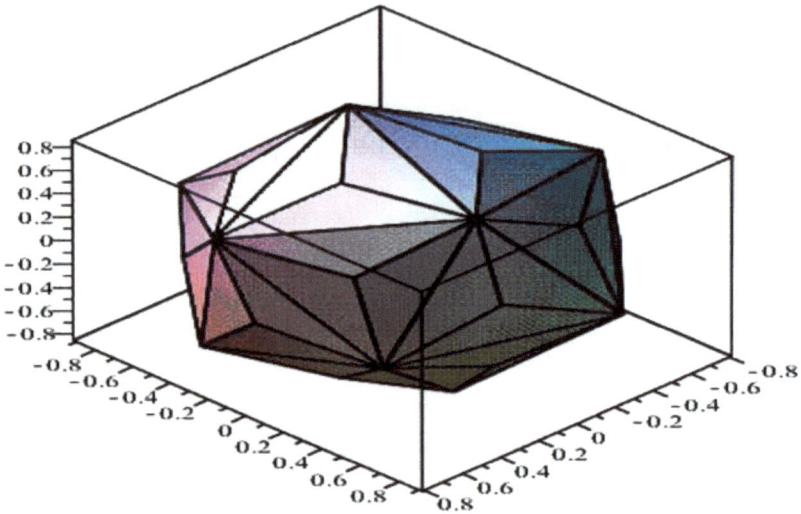

```
>JuliaSet:= proc(a,b)

 local z1, z2, z1s, z2s,m;

 (z1, z2): = (a,b):

 z1s:= z1^2:

 z2s: = z2^2;
```

**for m to 30 while z1s+z2s < 4 do**

  **(z1, z2):= (z1s-z2s, 2\*z1\*z2) + (0, 0.75);**

  **z1s:=z1^2;**

  **z2s:= z2^2;**

 **end do;**

**m;**

**end proc:**

>

*densityplot(JuliaSet, - 1.5 ..1.5, - 1.4 ..1.4, colorstyle = HUE, grid = [ 150, 150], style
    = patchnogrid, axes = none)*

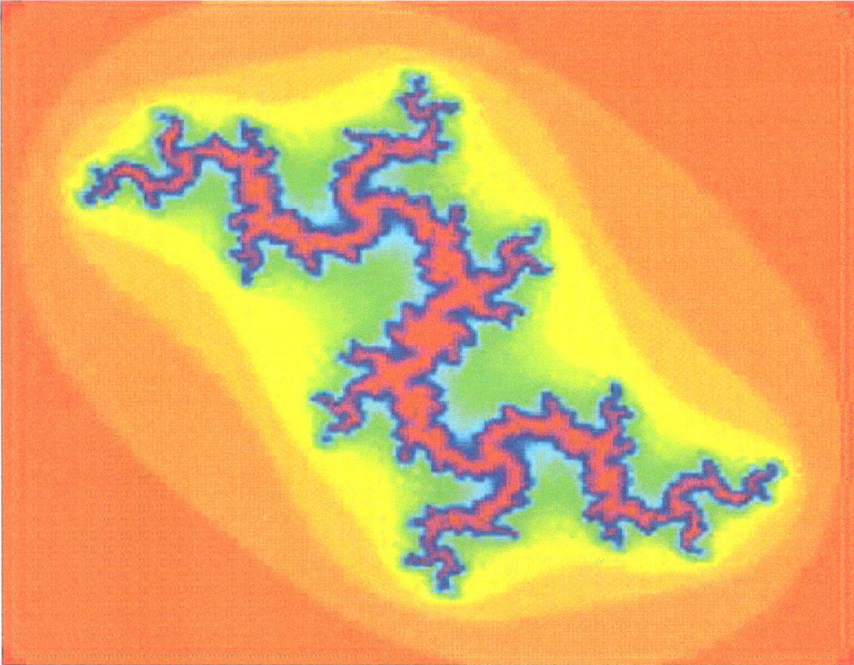

> *densityplot*( sin(*x y*), *x* = $-\pi$ ..$\pi$, *y* = $-\pi$ ..$\pi$, *axes* = *boxed, colorstyle* = *HUE*)

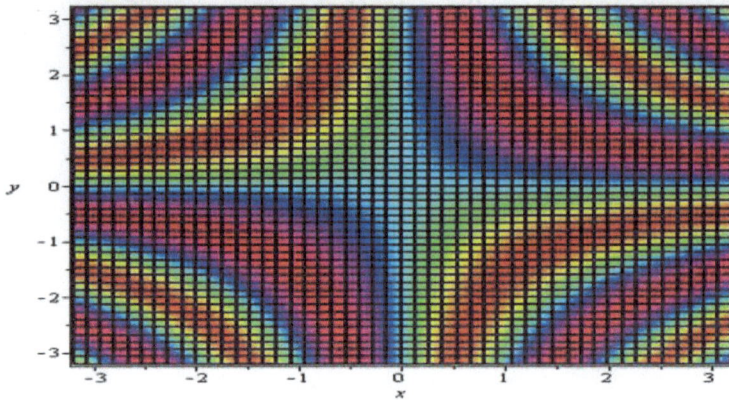

$$> densityplot\left( \exp\left( -\frac{x}{y + \text{Pi}} \right) \cdot \sin(x\,y), x = -\pi..\pi, y = -\pi..\pi, axes = boxed, colorstyle = HUE \right)$$

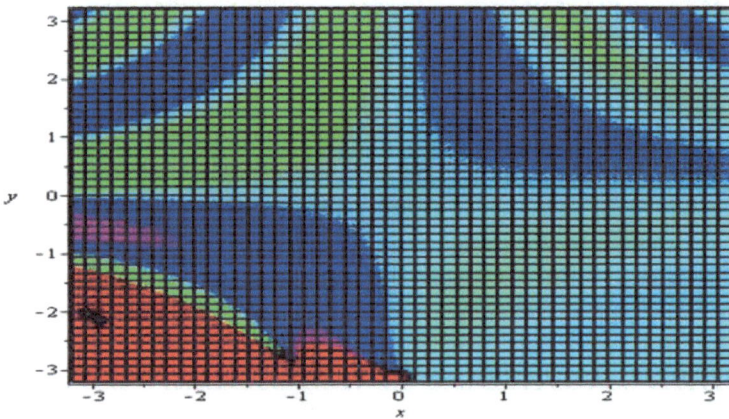

$$> densityplot(\sin(\text{Pi}\cdot x + y), x = -1..1, y = -1..1)$$

> $densityplot\left( \exp\left( \dfrac{x}{y} \right) \cdot \cos(\text{Pi} \cdot x - y) \cdot \sin(\text{Pi} \cdot x + y), x = -3 ..3, y = -3 ..3 \right)$

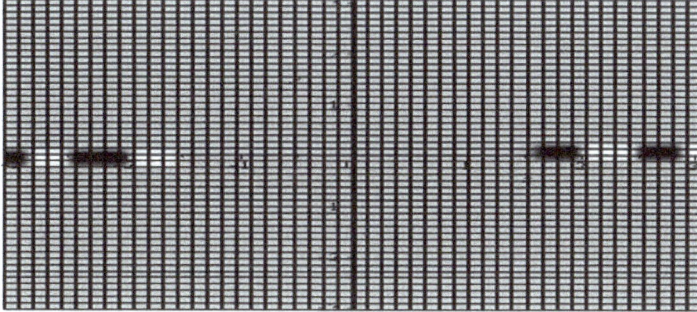

> $SpaceCurve\left( \langle e^{-t}\cos(t), e^{-t}\sin(t) \rangle, t = 4 ..8 \right)$

$$SpaceCurve\left( \begin{bmatrix} e^{-t}\cos(t) \\ e^{-t}\sin(t) \end{bmatrix}, t = 4 ..8 \right)$$

> $SpaceCurve(\langle \cos(t), \sin(t), t \rangle, t = 1 ..9)$

$$SpaceCurve\left( \begin{bmatrix} \cos(t) \\ \sin(t) \\ t \end{bmatrix}, t = 1 ..9 \right)$$

> $with(VectorCalculus) :$

> $SpaceCurve(\langle e^{-t}\cos(t), e^{-t}\sin(t) \rangle, t = -5 ..5)$

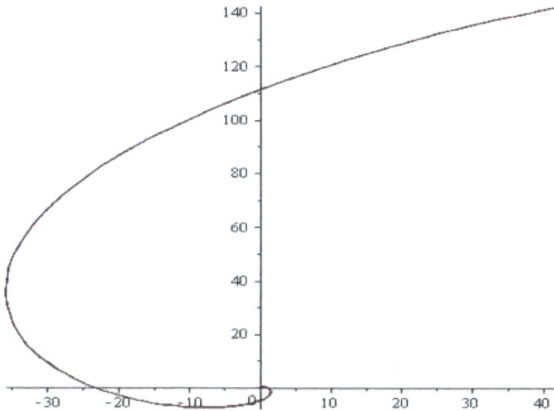

> $SpaceCurve(\langle\cos(t), \sin(t), t\rangle, t = 1 ..9)$

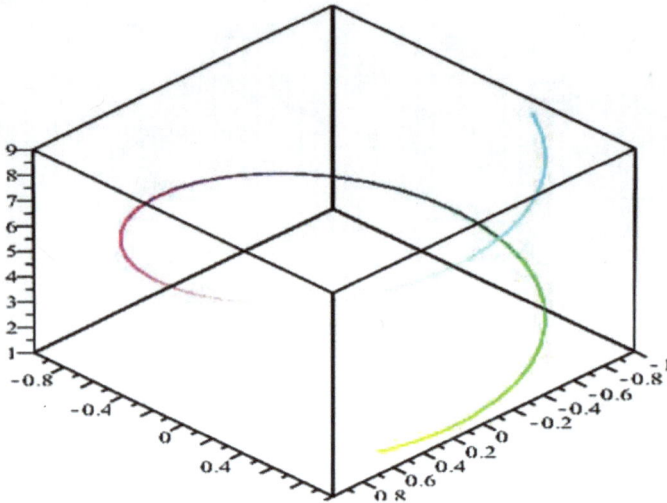

## N-order Nuclear Reactor Process

*with(ODETools)* :

> *with(plots)* :

> $eqn1 := diff(N(t), t) = \dfrac{(k-1)\cdot N(t)}{l} - \dfrac{beta\cdot N(t)}{l} + sum(lambda[i]\cdot r[i](t), i = 1 ..m);$

$$eqn1 := \frac{d}{dt} N(t) = \frac{(k-1)\,N(t)}{l} - \frac{\beta\,N(t)}{l} + \sum_{i=1}^{m} \lambda_i\, r_i(t)$$

>

$$eqn2 := diff(r[i](t), t) = \frac{beta\cdot N(t)}{l} - lambda[i]\cdot r[i](t);$$

$$eqn2 := \frac{d}{dt}\, r_i(t) = \frac{\beta\,N(t)}{l} - \lambda_i\, r_i(t)$$

> *wih(PDETools, dchange)* :

> $chngV1 := \{N(t) = sum(N[j]\cdot\exp(omega[j](t)), j = 0 ..m)\};$

$$chngV1 := \left\{ N(t) = \sum_{j=0}^{m} N_j e^{\omega_j(t)} \right\}$$

```
> chngV2 := {r[i](t) = sum(r[i,j]·exp(omega[j](t))), j = 0 ..m)};
```

$$chngV2 := \left\{ r_i(t) = \sum_{j=0}^{m} r_{i,j} e^{\omega_j(t)} \right\}$$

```
> P1 := op(factor(combine(expand(chngV1), power)))
```

$$P1 := N(t) = \sum_{j=0}^{m} N_j e^{\omega_j(t)}$$

```
> P2 := op(factor(combine(expand(chngV2), power)))
```

$$P2 := r_i(t) = \sum_{j=0}^{m} r_{i,j} e^{\omega_j(t)}$$

```
> evalf({eqn1, eqn2});
```

$$\left\{ \frac{d}{dt} N(t) = \frac{(k - 1.) N(t)}{l} - \frac{1.\beta N(t)}{l} + \sum_{i=1}^{m} \lambda_i r_i(t), \; \frac{d}{dt} r_i(t) = \frac{\beta N(t)}{l} - 1.\lambda_i r_i(t) \right\}$$

```
> subs([P1, P2], [eqn1, eqn2]);
```

$$\left[ \frac{\partial}{\partial t} \left( \sum_{j=0}^{m} N_j e^{\omega_j(t)} \right) = \frac{(k-1) \left( \sum_{j=0}^{m} N_j e^{\omega_j(t)} \right)}{l} - \frac{\beta \left( \sum_{j=0}^{m} N_j e^{\omega_j(t)} \right)}{l} + \sum_{i=1}^{m} \lambda_i \left( \sum_{j=0}^{m} r_{i,j} e^{\omega_j(t)} \right), \right.$$

$$\left. \frac{\partial}{\partial t} \left( \sum_{j=0}^{m} r_{i,j} e^{\omega_j(t)} \right) = \frac{\beta \left( \sum_{j=0}^{m} N_j e^{\omega_j(t)} \right)}{l} - \lambda_i \left( \sum_{j=0}^{m} r_{i,j} e^{\omega_j(t)} \right) \right]$$

```
> p := simplify(%);
```

$$p := \left[\sum_{j=0}^{m} N_j \left(\frac{d}{dt}\,\omega_j(t)\right) e^{\omega_j(t)} = \right.$$

$$-\frac{\beta\left(\sum_{i=0}^{m} N_i e^{\omega_i(t)}\right)-\left(\sum_{i=0}^{m} N_i e^{\omega_i(t)}\right)k-\left(\sum_{i=1}^{m}\lambda_i\left(\sum_{j=0}^{m} r_{i,j} e^{\omega_j(t)}\right)\right)l+\sum_{i=0}^{m} N_i e^{\omega_i(t)}}{l},\sum_{j=0}^{m}$$

$$\left. r_{i,j}\left(\frac{d}{dt}\,\omega_j(t)\right) e^{\omega_j(t)} = -\frac{\lambda_i\left(\sum_{j=0}^{m} r_{i,j} e^{\omega_j(t)}\right)l-\beta\left(\sum_{j=0}^{m} N_j e^{\omega_j(t)}\right)}{l}\right]$$

> *map( simplify,* **(28)**, *'assume = nonnegative'* )

$$\left[\sum_{j=0}^{m} N_j \left(\frac{d}{dt}\,\omega_j(t)\right) e^{\omega_j(t)} = \right.$$

$$-\frac{\beta\left(\sum_{i=0}^{m} N_i e^{\omega_i(t)}\right)-\left(\sum_{i=0}^{m} N_i e^{\omega_i(t)}\right)k-\left(\sum_{i=1}^{m}\lambda_i\left(\sum_{j=0}^{m} r_{i,j} e^{\omega_j(t)}\right)\right)l+\sum_{i=0}^{m} N_i e^{\omega_i(t)}}{l},\sum_{j=0}^{m}$$

$$\left. r_{i,j}\left(\frac{d}{dt}\,\omega_j(t)\right) e^{\omega_j(t)} = -\frac{\lambda_i\left(\sum_{j=0}^{m} r_{i,j} e^{\omega_j(t)}\right)l-\beta\left(\sum_{j=0}^{m} N_j e^{\omega_j(t)}\right)}{l}\right]$$

*Examples on image processing*

> *with( ImageTools )* :

> $img1 := Create\left(100, 200, (r, c) \rightarrow evalf\left(0.5\sin\left(\frac{r}{50}\right) + 0.5\sin\left(\frac{c}{30}\right)\right)\right)$ :

> $img2 := Complement( img1 )$ :

> *View( img1 )*

> *View( [ img1, img2 ] )*

> $PlotHistogram(img1, 100)$

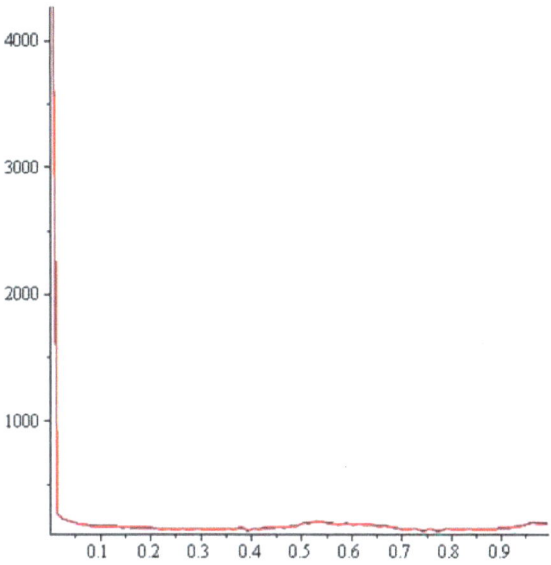

> $PlotHistogram(img2, 100, autorange)$

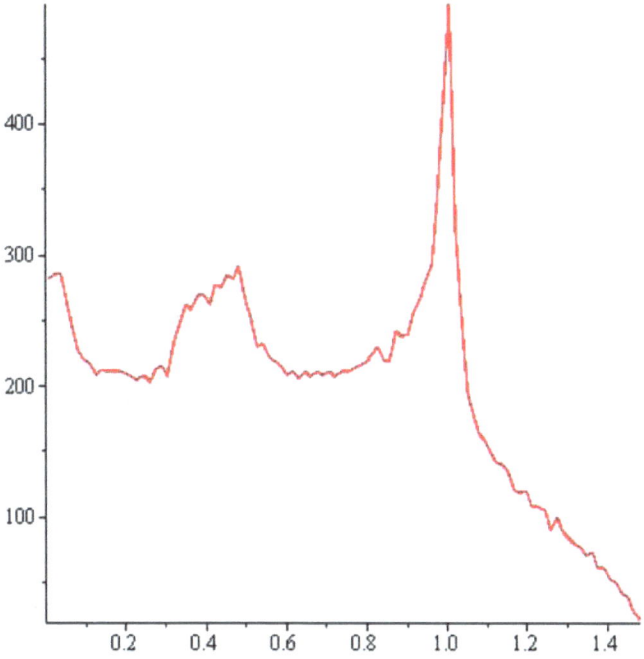

> *PlotHistogram*(*img2, autorange, normalized*)

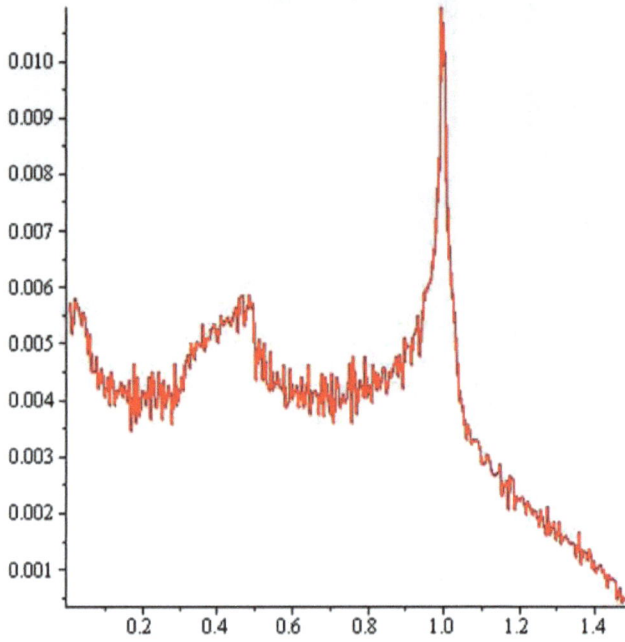

> *PlotHistogram*(*img1, autorange*)

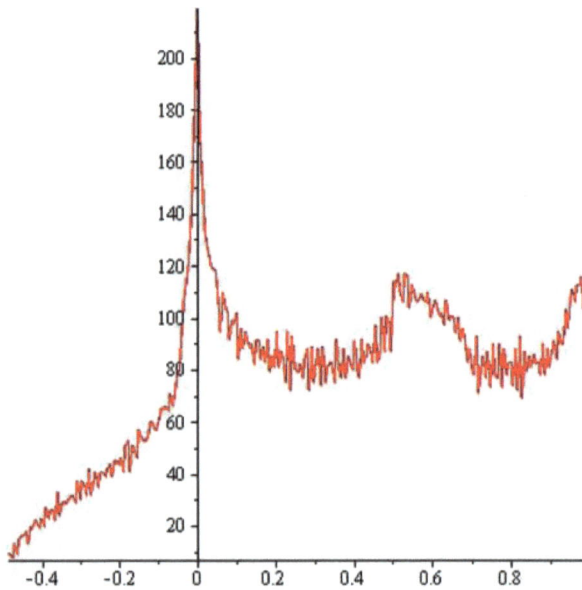

> *PlotHistogram*(*img2*, *range* = 0 ..0.5)

## Example for fitting experiments

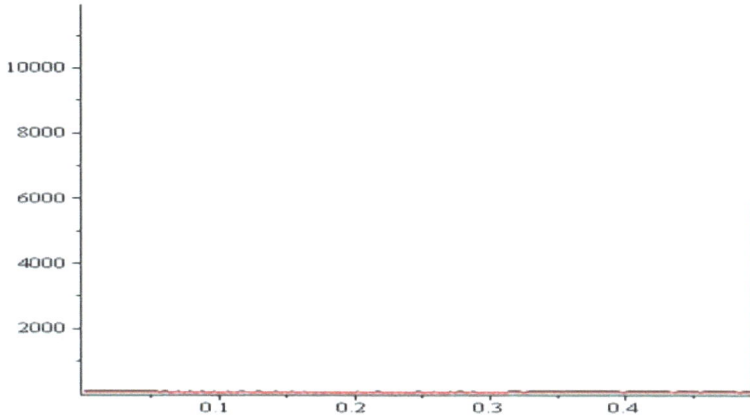

> *restart*

> *with*(*Statistics*) :

> *X* := *Vector*([1, 2, 3, 4, 5, 6], *datatype* = *float*) :

> *Y* := *Vector*([2, 5.6, 8.2, 20.5, 40.0, 95.0], *datatype* = *float*) :

> *ExponentialFit*(*X*, *Y*, *v*)

$1.01888654495804\, e^{0.746236510177018\, v}$

> *W* := *Vector*([1, 1, 1, 2, 5, 5], *datatype* = *float*) :

> *ExponentialFit*(*X*, *Y*, *weights* = *W*)

$$\begin{bmatrix} 0.989482297469512 \\ 0.752656239139387 \end{bmatrix}$$

> *LinearFit*([1, *t*, *t*$^2$], *X*, *Y*, *t*)

$24.6000000000000 - 23.9892857142857\, t + 5.79642857142857\, t^2$

$> LinearFit(a + bt + ct^2, X, Y, t)$

$24.6000000000000 - 23.9892857142857\,t + 5.79642857142857\,t^2$

Consider now an experiment where quantities $x$, $y$ and $z$ are quantities influencing a quantity $w$ according to an approximate relationship

$$w = ax + \frac{bx^2}{y} + cyz$$

with unknown parameters $a$, $b$, and $c$. Six data points are given by the following matrix, with respective columns for $x$, $y$, $z$, and $w$.

$>$
$ExperimentalData := \langle\langle 1, 1, 1, 2, 2, 2\rangle\langle 1, 2, 3, 1, 2, 3\rangle\langle 1, 2, 3, 4, 5, 6\rangle\langle 0.531, 0.341, 0.163, 0.641, 0.713, -0.040\rangle\rangle$

$$ExperimentalData := \begin{bmatrix} 1 & 1 & 1 & 0.531 \\ 1 & 2 & 2 & 0.341 \\ 1 & 3 & 3 & 0.163 \\ 2 & 1 & 4 & 0.641 \\ 2 & 2 & 5 & 0.713 \\ 2 & 3 & 6 & -0.040 \end{bmatrix}$$

$> LinearFit\left(\left[x, \frac{x^2}{y}, yz\right], ExperimentalData, [x, y, z]\right)$

$$0.823072918385878\,x - \frac{0.167910114211606\,x^2}{y} - 0.0758022678386438\,yz$$

$> NonlinearFit(a + bv + e^{cv}, X, Y, v)$

$$2.15979247107424 - 1.22391291112346\,v + e^{0.766784080984173\,v}$$

$>$
$NonlinearFit\left(x^a + \frac{bx^2}{y} + cyz, ExperimentalData, [x, y, z], initialvalues = [a = 2, b = 1, c = 0], output = [leastsquaresfunction, residuals]\right)$

$$\left[ x^{1.14701973996968} - \frac{0.298041864889394\, x^2}{y} - 0.0982511893429762\, yz, \right.$$

$$[0.0727069457676300, 0.116974310183398, -0.146607992383251,$$

$$\left. -0.0116127470057686, -0.0770361532848388, 0.0886489085642805]\right]$$

# APPENDIX B

# Introduction to MapleSim Software

**Abstract:** MapleSim is a modelling environment for creating and simulating complex multi-domain physical systems. It allows building component diagrams that represent physical systems in the graphical form. MapleSim automatically generates model equations from the component diagrams using symbolic and numerical approaches and runs very highly accurate simulations.

MapleSim modelling environment combines components from different engineering domains such as mechanical, electrical, and multi-body for building and exploring realistic designs and for studying the system level.

**Keywords:** Maple, MapleSim, Monte Carlo Simulation, Numerical, Symbolic, Simulation

## INTERACTIONS

In MapleSim environment

- Models' system level can be easily assessed to demonstrate concepts such as parameter optimization, sensitivity analyses, and interactions.

- Mathematical equations can be defined for new components from the first principle.
- Simulation can be carried out to investigate a much larger result of conditions that is possible. By testing with hardware alone, with no risk of damage to the equipment and for less cost.
- Allows export from MapleSim to C code, simulation, Labview, and other tools where it can be incorporated with a physical prototype.

### Features in MapleSim

- MapleSim have facilities for visualization in 3D and animation of multibody systems, full playback, and cameral control in 3D visualization.
- Interface and modelling: It contains drag-and-drop block diagrams in modelling environment, modelling diagrams, and 3-D model construction of multibody systems, data import, and export.
- Block Library: MapleSim contains both physical component and signal-flow blocks. The physical component blocks have different formalities for many domains.

**Benjamin Oyediran Oyelami**
**All rights reserved-© 2024 Bentham Science Publishers**

- Analysis and documentation: extract, view, and manipulation of the system equations for a model l, and parameter optimization. Simulation and parameter swaps including related files in a MapleSim model for easy documentation management and sharing.

Linear, nonlinear, continuous and discrete, SISO, MIMOS and hybrid systems parameter set managing and deployment to popular platforms from Mathword. MapleSim Connect can connect with Simulink.

## B1. Design of Simulation using the MapleSim

### B1.1 Code Generation

Code generation can handle all systems modeled in MapleSim, including hybrid systems with defined signal input (RealInput) and signal output (RealOutput) ports (MapleSim).

The source code in MapleSim is designed to interface with Maple, in the sample code; all inputs are assigned the value of 0. For more information about the available Code Generation command, see the GetCompiledProc help topic in MapleSim.

### C Code Generation

For C code generation, select the attachment of the generation of code from the MapleSim.

Step 1: Subsystem Selection

Click the button:

| Load Selected Subsystem |

Step 2: Inputs/Outputs and Parameter Management

### Inputs:

| | Input Variables | Change Row |
|---|---|---|
| 1 | | |

## Outputs:

| Toggle Export Column |
|---|

| | Output Variables | Export | Change Row |
|---|---|---|---|
| 1 | | | |
| 2 | ` Main.'output 1'.T `(t) | "X" | |
| 3 | ` Main.output2.T `(t) | "X" | |
| 4 | ` Main.output3.T `(t) | "X" | |

☐ Add an additional output port for subsystem state variables

## Parameters:

Click:

| Toggle Export Column |
|---|

Then, the parameters used in the model would be generated as:

| | Parameters | Value | Export | Change Row |
|---|---|---|---|---|
| 1 | | | "X" | |
| 2 | HC1_C | 15. | "X" | |
| 3 | HC2_C | 15. | "X" | |
| 4 | HC3_C | 15. | "X" | |
| 5 | TC1_G | 0.1e2 | "X" | |
| 6 | TC2_G | 8. | "X" | |

The C code for the modeling program can be generated using various solvers by selecting optimization optional and the max mean projection iteration

## Step 3: C Code Generation Options

## Solver Options:

Fixed step solver:   ◯ Euler   ◯ RK2   ◯ RK3   ◉ RK4   ◯ Implicit Euler

## Optimization Options:

Level of code optimization (0=None, 3=Full):

0          1          2          3

## Constraint Handling Options:

Maximum number of projection iterations:  3

Error tolerance:   0.1e-4

☑ Apply projection during event iterations

## Event Handling Options:

Maximum number of event iterations:  10

Width of event hysteresis band:  0.1e-9

## Baumgarte Constraint Stabilization:

☐ Apply Baumgarte constraint stabilization    ☑ Export Baumgarte parameters

Alpha:  10

Beta:  2

## Step 4: Generate C Code

Target directory:

```
C:\Users\Prof B O Oyelami
```
[ Browse ]

C-File:

```
MsimModel
```

## Click to generate the C code:

[ Generate C code ]

## Step 5: View C Code

```
/**************************************************
 * Automatically generated by Maple.
 * Created On: Fri Jun 12 04:49:35 2015.
 **************************************************/
#ifdef WMI_WINNT
#define EXP __declspec(dllexport)
#else
#ifdef X86_64_WINDOWS
#define EXP __declspec(dllexport)
#else
#define EXP
#endif
#endif
#include <stdlib.h>
#include <stdio.h>
#include <math.h>
#ifdef FROM_MAPLE
#include <mplshlib.h>
static MKernelVector kv;
EXP ALGEB M_DECL SetKernelVector(MKernelVector kv_in, ALGEB args) { kv=kv_in; ret
#else
#ifdef WMI_WINNT
#define M_DECL __stdcall
#else
#define M_DECL
```

## Monte Carlo Simulation

Author : Benjamin O Oyelami

Date:14 October 2022

---

## Model Description

Monte Carlo simulation (MCS) can be made on a MapleSim model. To generate MCS, you define a random distribution for a parameter and you can run multiple simulations using this distribution. Note that the properties that are plotted are defined by the probes in the MapleSim model.

## Monte-Carlo Simulation

To start, click **Load System**.

> Load System

## Parameter Distribution

 Select the parameter you want to vary, and then choose an appropriate distribution and distribution parameters.

Parameter　　　HC3.C ▼

Nominal Value　　　15

Distribution　　Uniform ▼

Choose　　the
parameter　　a　　Help
and　b,　the

uniform
distribution as
follows:

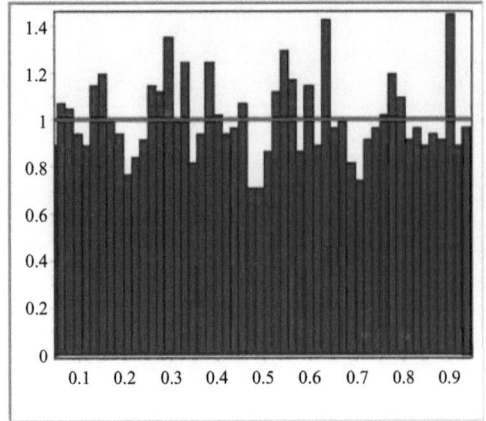

a                    | 0 |

b                    | 1 |

## Monte-Carlo Simulation

Enter the number of simulation runs and the number of bins in the simulation, and then click the **Run Simulation** button to create and display the simulation plots.

Number of simulations run (including nominal value) | 6 |

To plot variation, in the boxes click:

| Run Simulation |   for | All probes  ▼ |            ☐ Plot variances in boxes

Click Run simulation for the given problem and number of bins and the probe plots are displayed above:                                   | 12 |   **Number of bins**

**Note:** The blue line corresponds to the nominal values.

## Data Analysis

Specify a time value below (any float value between 0 and $tf$ s), choose an output variable in the list, and then click **Analyze Data**. Statistics quantities will be displayed on a data set of [5] points, with each point corresponding to one of the simulations, not including the nominal. More information on the quantities displayed and plotted; see Statistics in the Maple Help. The data on which the quantities are computed and plotted are stored as a list of Matrices in the variable _data . The first element corresponds to the nominal value (which is not used in the statistics). Select the desired sample of output variables and click analyze data and the statistics quantities displayed as follows:

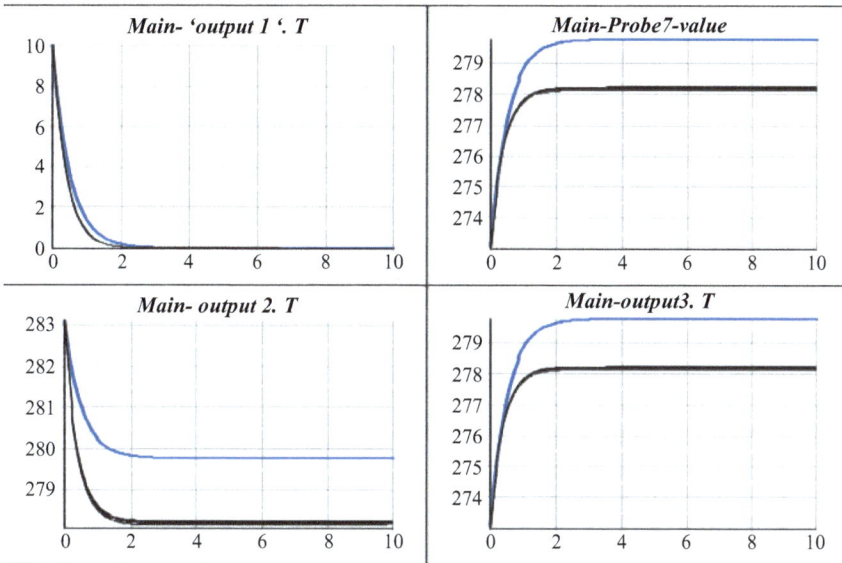

Sample Time          5                                    Analyze Data

Output Variable      Main.'output 1'.T  ▾

Statistics Quantities

```
0.000005
Skewness
-0.517357
Standard Deviation
0.000002
Variance
0.000000
Variation
0.084149
```

Select the type of plot you desire and click on the example,

Choose the Kernel/density plot and the plot displayed as follows:

KernelDensityPlot ▼

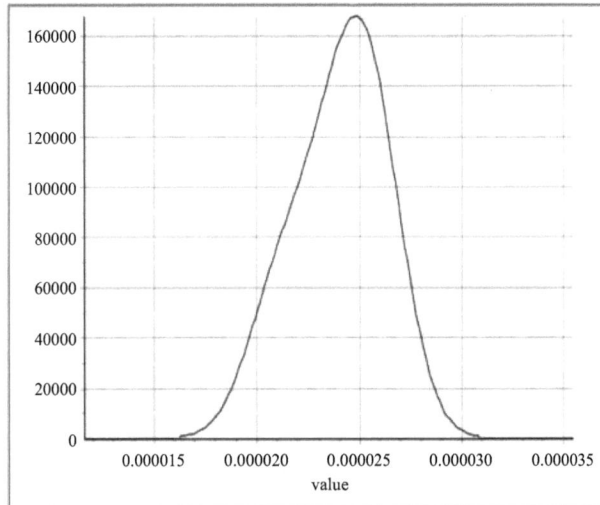

**Elastic Bearing**
**About this model:** This example shows a gearbox system where the housing does not directly connect to ground. The housing connects to a spring-damper system to allow for modeling of some the dynamics.

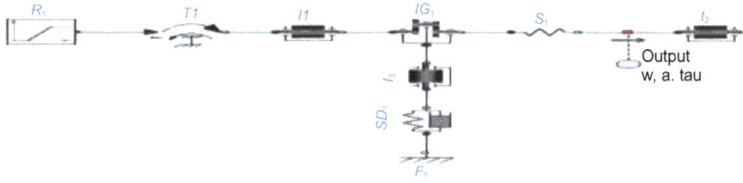

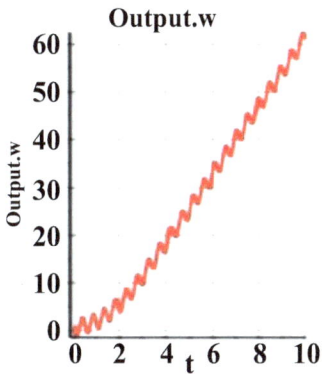

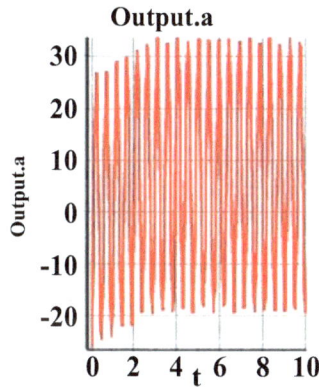

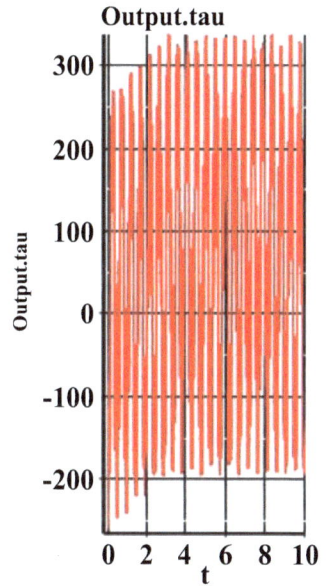

Save this worksheet in Maple and then save the **msim** file to which this worksheet is attached in MapleSim.

# SUBJECT INDEX

**Benjamin Oyediran Oyelami**
**All rights reserved-© 2024 Bentham Science Publishers**

www.ingramcontent.com/pod-product-compliance
Lightning Source LLC
Chambersburg PA
CBHW050813220326
41598CB00006B/196

* 9 7 8 9 8 1 5 3 1 3 8 8 8 *